# Siam Physics Congress 2017 (SPC 2017)

AA006323

Journal of Physics: Conference Series Volume 901

Rayong, Thailand
24 – 26 May 2017

ISBN: 978-1-5108-4927-3
ISSN: 1742-6588

**Printed from e-media with permission by:**

Curran Associates, Inc.
57 Morehouse Lane
Red Hook, NY 12571



Printed by Curran Associates, Inc. (2017)

For permission requests, please contact the Institute of Physics
at the address below.

Institute of Physics
Dirac House, Temple Back
Bristol BS1 6BE UK

Phone: 44 1 17 929 7481
Fax: 44 1 17 920 0979

techtracking@iop.org

**Additional copies of this publication are available from:**

Curran Associates, Inc.
57 Morehouse Lane
Red Hook, NY 12571 USA
Phone: 845-758-0400
Fax: 845-758-2633
Email: curran@proceedings.com
Web: www.proceedings.com

# TABLE OF CONTENTS

# Constraints on Dark Matter Annihilation by Synchrotron Emission based on Planck Data

**Chalit Muanglay, Maneenate Wechakama and Brandon K. Cantlay**
Department of Physics, Faculty of Science, Kasetsart University, Bangkok, Thailand
E-mail: chalit.ml@gmail.com, mwechakama@gmail.com

**Abstract.** Synchrotron emission can be a good probe for dark matter particles in the Milky Way. We have investigated the production of electrons and positrons in the Milky Way within the context of dark matter annihilation. Upper limits on the relevant cross-section are obtained by comparing synchrotron emission in the microwave bands with Planck data. According to our results, the dark matter annihilation cross-section into electron-positron pairs should not be higher than the canonical value for a thermal relic if the mass of the dark matter candidate is smaller than a few GeV. In addition, we also look for constraints on the inner slope of dark matter density profile in the Milky Way. Our results indicate that the inner slope of dark matter profile is between 1 to 1.5.

## 1. Introduction

Dark matter can be indirectly detected through the signatures of standard model particles produced by its annihilation or decay, see e.g. [1]. The recent results from indirect detection experiments in the solar neighbourhood have suggested the possibility that such a signature has been seen. In particular, the PAMELA experiment has pointed a significant excess of electrons and positrons above the expected smooth astrophysical background [2]. If these results are interpreted in terms of dark matter annihilation, then an abundant population of high-energy $e^{\pm}$ is being created everywhere in the Galactic dark matter halo, with the associated synchrotron emission in the Galactic magnetic field.

The present work focuses on the astrophysical signatures of dark matter annihilation into electron-positron pairs. We try to impose constraints on the relevant cross-section and on the dark matter density profile by comparing the predictions of an analytic model of synchrotron emission related to dark matter annihilation with a set of observational data obtained from Planck.

## 2. Method

### *2.1. Electron-positron propagation*

As in our previous work [3][4], we adopt a model-independent approach in which all the injected particles are created with the same initial energy $E_0$, of the order of the mass of the dark matter particle. Since electrons and positrons will be relativistic at the moment of their creation, they can efficiently lose their energy through different processes. We will use the Lorentz factor $\gamma$ to express the energy $E = \gamma m_e c^2$ of the annihilation products, where $m_e$ denotes the rest mass of electron, and $c$ is the speed of light.

The propagation of electrons and positrons through the interstellar medium (ISM) is determined by the diffusion-loss equation

$$\begin{aligned}\frac{\partial}{\partial t}\frac{\mathrm{d}n}{\mathrm{d}\gamma}(\mathbf{x},\gamma) &= \nabla\left[K(\mathbf{x},\gamma)\nabla\frac{\mathrm{d}n}{\mathrm{d}\gamma}(\mathbf{x},\gamma)\right] \\ &+ \frac{\partial}{\partial\gamma}\left[b(\mathbf{x},\gamma)\frac{\mathrm{d}n}{\mathrm{d}\gamma}(\mathbf{x},\gamma)\right] \\ &+ Q(\mathbf{x},\gamma). \end{aligned} \tag{1}$$

We assume a diffusion coefficient of the form $K(\gamma) = K_0\gamma^{\delta}$. The energy loss rate $b(\mathbf{x},\gamma)$ is a sum over the relevant physical processes, i.e. inverse Compton scattering, synchrotron radiation, Coulomb collisions, bremsstrahlung, and ionization. The source term $Q(\mathbf{x},\gamma)$ represents the instantaneous electron-positron injection rate.

Given enough time, the electron-positron population will approach a steady-state distribution. Considering a spherically-symmetric source, the spatial integral can be reduced to one dimension, and the electron-positron spectrum is finally given by the expression

$$\begin{aligned}\frac{\mathrm{d}n}{\mathrm{d}\gamma}(r,\gamma) = \frac{1}{b(\gamma)}\frac{\exp\left(-\frac{r^2}{2\Delta\lambda^2}\right)}{(2\pi r^2\Delta\lambda^2)^{1/2}} \times \Big\{ \int_{\gamma}^{\infty}\mathrm{d}\gamma_{\mathrm{s}}\int_0^{\infty}\mathrm{d}r_{\mathrm{s}}\, r_{\mathrm{s}}\, \exp\left(-\frac{r_{\mathrm{s}}^2}{2\Delta\lambda^2}\right) \\ \left[\exp\left(\frac{rr_{\mathrm{s}}}{\Delta\lambda^2}\right) - \exp\left(-\frac{rr_{\mathrm{s}}}{\Delta\lambda^2}\right)\right] Q(r_{\mathrm{s}},\gamma_s)\Big\}. \end{aligned} \tag{2}$$

### 2.2. *Source term*

The electrons and positrons in our model originate from the annihilation of dark matter particles. We consider a perfectly spherically-symmetric halo, the production rate is dictated by the dark matter number density and the annihilation rate into electron-positron pairs,

$$Q(r,\gamma) = n_{\mathrm{dm}}(r)\; n_{\mathrm{dm}^*}(r)\; \langle\sigma v\rangle_{e^\pm}\, \frac{\mathrm{d}N_{e^\pm}}{\mathrm{d}\gamma} \tag{3}$$

where $n_{\mathrm{dm}}$ and $n_{\mathrm{dm}^*}$ denote the number densities of dark matter particles and anti-particles, respectively, $\langle\sigma v\rangle_{e^\pm}$ is the thermal average of the annihilation cross-section times the dark matter relative velocity, and $\frac{\mathrm{d}N_{e^\pm}}{\mathrm{d}\gamma}$ is the injection spectrum of electrons and positrons in the final state. Assuming all electrons and positrons are injected with the same energy $\gamma_0 \sim m_{\mathrm{dm}}/m_{\mathrm{e}}$,

$$Q(r,\gamma) = Q_0(r)\;\delta(\gamma-\gamma_0) \tag{4}$$

where

$$Q_0(r) = 2\left[\frac{\rho_{\mathrm{dm}}(r)}{m_{\mathrm{dm}}}\right]^2 \langle\sigma v\rangle_{e^\pm} \tag{5}$$

is the local production rate per unit volume per unit time and $\delta(\gamma-\gamma_0)$ denotes a Dirac delta function.

For the dark matter density $\rho_{\mathrm{dm}}(r)$, we consider a perfectly spherically-symmetric halo described by a density profile of the form

$$\rho_{\mathrm{dm}}(r) = \frac{\rho_{\mathrm{s}}}{\left(\frac{r}{r_{\mathrm{s}}}\right)^{\alpha}\left(1+\frac{r}{r_{\mathrm{s}}}\right)^{3-\alpha}}, \tag{6}$$

where $r_{\mathrm{s}}$ and $\rho_{\mathrm{s}}$ denote a characteristic density and radius of the halo, respectively, and the $\alpha$ is the inner logarithmic slope of the density profile [5].

### *2.3. Surface brightness profile*

Once the electron-positron spectrum is computed, the emission coefficient[1] for photons of frequency $\nu$ is given by the integral

$$j_\nu(r,\nu) = \frac{1}{4\pi}\int_1^\infty \frac{\mathrm{d}n}{\mathrm{d}\gamma}(r,\gamma)\, l(\gamma,\nu)\, \mathrm{d}\gamma \tag{7}$$

of the electron-positron spectrum $\frac{\mathrm{d}n}{\mathrm{d}\gamma}(r,\gamma)$ times the specific luminosity $l(\gamma,\nu)$ emitted at frequency $\nu$ by a single electron or positron with Lorentz factor $\gamma$. The intensity from any given direction in the sky is simply the integral along the line of sight of the emission coefficient. Since we assume a spherically-symmetric source and boundary conditions, it will only depend on the angular separation $\theta$ with respect to the Galactic centre,

$$I_\nu(\theta,\nu) = \int_0^\infty j_\nu(r,\nu)\, \mathrm{d}s, \tag{8}$$

where $s$ represents the distance along the line of sight, and the radial distance $r$ to the centre of the Milky Way at any point along the ray is

$$r = \sqrt{x^2+y^2}, \tag{9}$$

with $x = s\sin\theta$, $y = s\cos\theta - R_\odot$, and $R_\odot = 8.5$ kpc (the distance of the Sun from the Galactic centre).

The contribution of synchrotron radiation, which dominates at low photon energies, can be estimated as (see e.g.[6])

$$l_{\mathrm{syn}}(\gamma,\nu) = \frac{\sqrt{3}\,q_{\mathrm{e}}^3 B}{m_{\mathrm{e}}c^2}\, R[\chi(\gamma)], \tag{10}$$

where $m_{\mathrm{e}}$ and $q_{\mathrm{e}}$ denote the electron mass and charge, respectively, $B$ is the intensity of the magnetic field, and the function $R(\chi)$ is defined as, e.g.[7],

$$R(\chi) \equiv 2\chi^2\left[K_{\frac{4}{3}}(\chi)K_{\frac{1}{3}}(\chi) - \frac{3}{5}\chi\left\{K_{\frac{4}{3}}^2(\chi) - K_{\frac{1}{3}}^2(\chi)\right\}\right]. \tag{11}$$

In this expression, $K$ refers to the modified Bessel function, and the normalized frequency

$$\chi \equiv \frac{\nu}{3\gamma^2\nu_c} \tag{12}$$

is expressed in terms of the cyclotron frequency

$$\nu_c \equiv \frac{q_e B}{2\pi m_e c}. \tag{13}$$

We calculate the electron-positron spectrum as described in expression (2), and then estimate the photon intensity according to expression (8).

### *2.4. Astrophysical parameters*

In our canonical model, the ISM is mainly composed of neutral hydrogen atoms ($X_{\mathrm{ion}} = 0$) with number density $\rho_{\mathrm{g}}/m_{\mathrm{p}} \sim 1$ cm$^{-3}$, and it is permeated by a tangled magnetic field whose intensity is $B \sim 6$ $\mu$G throughout the Galaxy. We use a dark matter density profile with $\alpha = 1$ [5], $r_{\mathrm{s}} = 17$ kpc and $\rho_{\mathrm{s}}c^2 = 0.35$ GeV cm$^{-3}$, consistent with dynamical models of the Milky Way. The virial mass of the Galaxy is thus $10^{12}$ M$_\odot$, and the local dark matter density is $\rho_{\mathrm{dm}}(r_\odot)\,c^2 = 0.3$ GeV cm$^{-3}$. For the diffusion coefficient, we consider the MED model discussed by [8]. We will also use a model of the ISRF [9] where the photon intensity is represented by grey-body components.

[1] Energy radiated per unit volume per unit frequency per unit time per unit solid angle.

### *2.5. Observational data*
In order to constrain the production of relativistic electrons and positrons in the Milky Way, we consider observations of the whole sky at different wavelengths by using data from 9 channels of Planck observation at microwave wavelengths, which dominated by synchrotron emission. We take the full-resolution coadded temperature maps for each of the 9 frequency bands (30 GHz, 44 GHz, 70 GHz, 100 GHz, 143 GHz 217 GHz, 353 GHz, 545 GHz and 857 GHz) [10].

## 3. Results
Once the emission from the Galactic disc and the most prominent point sources is excluded, the remaining spherically-averaged component can be used to place upper limits on the cross-section for dark matter annihilation into electron-positron pairs.

First of all, model intensities are computed according to the scheme described in Section 2.3. We consider the injection energy (i.e. the mass of the dark matter particle) as a free parameter and investigate values of the initial Lorentz factor $\gamma_0$ between $2 \times 10^3$ and $2 \times 10^7$. We have found that the dark matter annihilation cross-section into electron-positron pairs should not be higher than the canonical value for a thermal relic if the mass of the dark matter candidate is smaller than a few GeV.

By assuming a given value of the cross-section, one can constrain the value of $\alpha$ in expression (6) from the total intensity and the morphology of the observed surface brightness. We set the dark matter annihilation cross-section into electron-positron pairs to the value expected for a thermally-produced relic, $\langle \sigma v \rangle_{e^\pm} = 3 \times 10^{-26}$ cm$^3$ s$^{-1}$, and compute the value of $\alpha$ for which the predicted emission rises above the observed level. Our results indicate that the inner slope of dark matter profile $\alpha$ is between 1 to 1.5.

## Acknowledgments
This research is supported in part by the Graduate Program Scholarship from the Graduate School, Kasetsart University.

## References
[1] Bertone G 2010 *Particle Dark Matter* (Cambridge: Cambridge University Press)
[2] Adriani O, et al., PAMELA Collaboration 2009 *Nature.* **458** 607
[3] Wechakama M, and Ascasibar Y 2011 *MNRAS* **413** 1991
[4] Wechakama M, and Ascasibar Y 2014 *MNRAS* **439** 566
[5] Navarro J F, Frenk C S, and White S D M 1996 *ApJ* **462** 563
[6] Sarazin C L 1999 *ApJ* **520** 529
[7] Ghisellini G, Guilbert P W, and Svensson R 1988 *ApJL* **334** 5
[8] Donato F, et al. 2004 *Phy. Rew. D.* **69** 063501
[9] Cirelli M, and Panci P 2009 *Nuclear Physics B* **821** 399
[10] Planck Collaboration 2016 *Astronomy & Astrophysics* **594** A1

# Effect of a modified sinusoidal forcing on spiral wave in a simulated reaction-diffusion system.

**Kritsana Khaothong[1], Malee Sutthiopad[1], Nakorn Kumchaiseemak[1], Jiraporn Luengviriya[2], Jarin Kanchanawarin[1], Stefan C. Müller[3] and Chaiya Luengviriya[1,*]**

[1] Department of Physics, Kasetsart University 50 Phaholyothin Road, Jatujak, Bangkok 10900, Thailand

[2] Lasers and Optics Research Group, Department of Industrial Physics and Medical Instrumentation, King Mongkut's University of Technology North Bangkok, 1518 Pibulsongkram Road, Bangkok 10800, Thailand

[3] Institute of Experimental Physics, Otto-von-Guericke University Magdeburg, Universitätsplatz 2, D-39106 Magdeburg, Germany

* E-mail : fscicyl@ku.ac.th

**Abstract**. Spiral waves are often found in excitable media. In the hearts, they are abnormal forms of action potential propagation. Under an external forcing, the spiral waves drift and are subsequently terminated at the boundary. Spiral waves can be studied in simulations using a discrete reaction-diffusion system; thereby the time step must not exceed a numerical stability limit ($t_s$). In this article, we present the dynamics of spiral waves in a simulated system under an external forcing as a modified sinusoidal function of time. The spiral waves are forced to drift along a straight line with a velocity and an angle depending on the time step. An optimization study provides the optimal time step of $0.2t_s$, where further reductions of the time step do not alter the drifting of the spiral waves.

## 1. Introduction

Spiral waves are often found in excitable media, e.g., cell aggregation in slime mold colonies [1], during CO-oxidation on a platinum surface [2], electrical wave propagation in cardiac tissues [3], and the Belousov-Zabotinsky (BZ) reaction [4]. Action potential propagation in forms of spiral waves in heart tissues related to some cardiac arrhythmias and can lead to sudden cardiac death. Under an external forcing, the spiral waves drift and are subsequently terminated at the boundary of the medium. Drifting of spiral waves can be induced by an applied electric field and the drifting velocity and angle of the spiral waves increase with the field strength [5]. Under a sufficient strong field, the spiral waves are forced to hit the boundary and eventually annihilated [6].

Dynamics of spiral waves and the effect of forcing in excitable media are often studied in a discrete reaction-diffusion system [7,8]. To avoid errors from calculations, the time step used in the simulations must not exceed a numerical stability limit ($t_s$) [9]. In this article, we present the effect of the time step on the dynamics of spiral waves in a simulated system.

## 2. Methods

In our numerical simulations, we use the two-variable Oregonator model [7] to describe the dynamics of the activator u and inhibitor v in excitable media. The advection terms for both $u$ and $v$ account for electric field for sinusoidal function E applied in x-direction.

$$\frac{\partial u}{\partial t} = \frac{1}{\varepsilon}\left(u - u^2 - fv\frac{u-q}{u+q}\right) + D_u\nabla^2 u - M_u E\frac{\partial u}{\partial x}$$
$$\frac{\partial v}{\partial t} = u - v + D_v\nabla^2 v - M_v E\frac{\partial v}{\partial x} \tag{1}$$

As in References [8], the parameters are chosen as $\varepsilon$ = 0.01, $q$ = 0.002, $f$ = 1.4, the diffusion coefficients $D_u$ = 1.0 and $D_v$ = 0.6, and the ionic mobilities $M_u$ = -1.0 and $M_v$ = 2.0. In the absence of an electric field, the tip of a free spiral wave rotates around a circular core (diameter = 0.9 s.u.).

The simulations are performed using the explicit Euler method with 9-point approximation of the two-dimensional Laplacian operator and a centered-space approximation of the gradient term. We use the uniform grid space $\Delta x = \Delta y$ = 0.1 system unit (s.u.). The time step $\Delta t$ is varied between 0.01 - 0.9 $t_s$, where the numerical stability limit $t_s = (3/8)\,(\Delta x)^2$ [9]. The dimensionless size of the system is 40 x 40 s.u. (400 x 400 grid points). To create a spiral wave, a planer wave is triggered by setting 5-grid-point strip at an edge of the medium to an excited state. The wave front is allowed to propagate into the middle of the medium, before half of the medium is reset to an excitable state. Then we apply the forcing $E$ as a sinusoidal function of time t

$$E = \frac{E_0}{2}\sin\left(\frac{2\pi t}{T}\right) + \frac{E_0}{2} \tag{2}$$

where the amplitude $E_0$ and the period ($T$) are kept constant at 0.4, 0.3 t.u. We observe the dynamics of the spiral wave both with and without the applied forcing and search for the optimal time step i.e., the largest time step that further reduction of the time step does not change the spiral wave dynamics.

## 3. Results

In the absence of the applied forcing, the free spiral wave, as shown in figure 1a, rotates around a small circular circle. When the time step $\Delta t$ is decreased, the wavelength is approximately constant while the wave period $T_w$ are increased [figure 1 (b)] but the wave velocity $v_w$ decreased as shown in figures 1(b) and 1(c). The results imply that the optimal time step is $\Delta t$ = 0.4 $t_s$ since the smaller time steps give very similar results.

The motion of a spiral wave under an applied sinusoidal forcing with the time step $\Delta t$ = 0.8 $t_s$ is shown in figure 2. The forcing causes the spiral wave whose tip is at the middle of the system [figure 2(a)], to drift along a straight path to the left with an angle with respect to the forcing direction pointing to the right [figures 2(a)-2(d)]. It is subsequently terminated at the boundary of the medium [figure 2(e)].

Figure 3 illustrates the detailed analysis of all simulations with an applied sinusoidal forcing. A graph of modified sinusoidal forcing (see equation 2) plotted against time is shown in figure 3(a). Even though the spiral wave drift linearly under the applied forcing, the drifting velocity and angle depend on the time step as shown in figures 3(b) and 3(c). When the time step $\Delta t$ *is* decreased until $\Delta t$ = 0.2 $t_s$, both the drifting velocity and angle increase. However, reductions to smaller time steps do not change the drifting velocity and angle. Thus, the optimal time step of $\Delta t$ = 0.2 $t_s$ for the simulations with sinusoidal forcing.

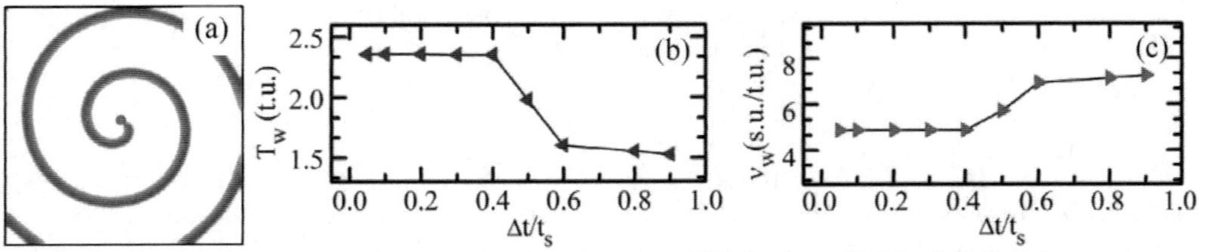


**Figure 1.** Dynamics of a spiral wave in the absence of applied forcing. (a) The free spiral wave rotates around a small circle (the filled circle) with (b) wave period $T_w$ and (c) wave velocity $v_w$ as a function of the time step $\Delta t$.

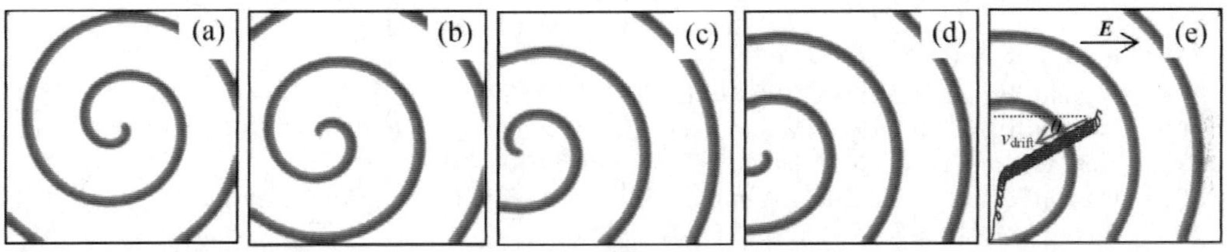


**Figure 2.** Sequential images of a spiral wave under a sinusoidal forcing. The spiral tip (a) initially located at the middle of the system (b-d) is gradually drifts to the left with an angle (e) until hits the boundary where the spiral wave is eliminated. The blue line indicates the drifting trajectory of the spiral tip. When the tip is located near the boundary, it drifts approximately with an angle $\theta_{drift}$ to the direction of forcing.

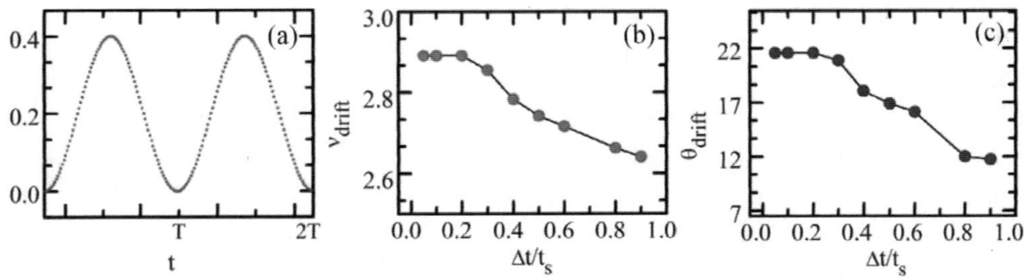


**Figure 3.** Effect of time step on the spiral wave under a sinusoidal forcing. (a) The sinusoidal forcing E vs time t. The spiral tip drifts with (b) velocity $v_{drift}$ and (c) angle $\theta_{drift}$ depending of the time step $\Delta t$.

## 4. Discussion and Conclusion

We have presented an investigation on the effect of the time step on the dynamics of spiral waves at different time step in a simulated system. Both the motion of free spiral waves (in the absence of forcing) and the drifting of spiral waves under an applied sinusoidal forcing depend on the time step. The optimal time step in the case of time varying forcing ($0.2t_s$) which is smaller than that for the simulations of free spiral waves ($0.4t_s$). This comes from the fact that the applied time-dependent forcing is discretized and becomes a step-wise function which depends on the time step. Therefore, simulations of spiral waves under an applied forcing as a function of time should be performed with a fine time step to obtain promising results for comparison and confirmation of experimental results.

It has been shown that the lifetime of tachycardia is extended by pinning of spiral waves to anatomical obstacles like veins and scars [3] and the elimination of such pinned spiral waves become a tough task especially for large obstacles [10]. Even though we show that the sinusoidal forcing induce a linear drift of spiral waves very similar to the case of constant forcing, it is interesting to investigate that whether an application of time varying forcing as in this article can improve the success of unpinning of spiral waves.

## 5. Acknowledgment

We thank the Department of Physics, Faculty of Science, the Research and Development Institute (KURDI), the Center for Advanced Studies of Industrial Technology, and the Graduate School, Kasetsart University for financial support.

## 6. References

[1] Siegert F and Weijer C J 1989 *J. Cell Sci.* **93** 325
[2] Nettesheim S, Oertzen A von, Rotermund H. H and Ertl G 1993 *J. Chem. Phys.* **98** 9977
[3] Davidenko J M, Pertsov A M, Salomonz R, Baxter W and Jalife J 1992 *Nature.* **335** 349
[4] Winfree A T 1972 *Science.* **175** 634
[5] Schmidt B and Müller S C 1997 *Phys. Rev. E* **55** 4390
[6] Luengviriya J, Sutthiopad M and Phantu M 2014 *Phy. Rev. E* **90** 052919
[7] Field R. J and Noyes R M 1974 *J. Chem. Phys.* **60** 1877
[8] Jahnke W, Skaggs W E and Winfree A T 1989 *J. Phy. Chem.* **93** 740
[9] Dowle M, Mantel R. M and Barkley D 1997 *Int. J. Bif. Chaos.* **7** 2529
[10] Sutthiopad M, Luengviriya J, Porjai P, Tomapatanaget B, Müller S.C and Luengviriya C 2014 *Phy. Rev. E* **89** 052902

# Gibbs Energy Additivity Approaches in Estimation of Dynamic Viscosities of n-Alkane-1-ol

**S. Phankosol[1*] and K. Krisnangkura[2]**

[1]Faculty of Science and Technology, Bansomdejchaopraya Rajabhat University, Bangkok, 10600, Thailand
[2] School of Bioresources and Technology, King Mongkut's University of Technology Thonburi (Bangkhuntien), Bangkok, 10150, Thailand

*Corresponding's author: s.phankosol@gmail.com

**Abstract**. Alcohols are solvents for organic and inorganic substances. Dynamic viscosity of liquid is important transport properties. In this study models for estimating n-alkan-1-ol dynamic viscosities are correlated to the Martin's rule of free energy additivity. Data available in literatures are used to validate and support the proposed equations. The dynamic viscosities of n-alkan-1-ol can be easily estimated from its carbon numbers ($n_c$) and temperatures ($T$). The bias, average absolute deviation and coefficient of determination ($R^2$) in estimating of n-alkan-1-ol are -0.17%, 1.73% and 0.999, respectively. The dynamic viscosities outside temperature between 288.15 and 363.15 K may be possibly estimated by this model but accuracy may be lower.

## 1. Introduction

Alcohols are solvents for organic and inorganic substances. Recently, alcohols have been proposed as an alternatives as oxygenated fuel. Oxygenated fuel is used to help fuel burn more efficiently and cut down on some types of atmospheric pollution.[1] Viscosity of liquid is important for equipment design, heat and mass transfer. There were many published mathematical models for prediction of viscosity of the liquids, including vegetable oils. The Andrade equation (Eq.1) was a good equation that correlates dynamic viscosity ($\eta$) to the free energy of viscous flow ($\Delta$G) and absolute temperature ($T$).[2, 3]

$$\eta = A_0 e^{-\Delta G / RT} \qquad (1)$$

or

$$\ln \eta = A + \frac{B}{T} \qquad (2)$$

where $A_0$ and $A$ are constant and $\ln A_0$, respectively.
$B$ is $\Delta G/RT$
R is universal gas constant.

However, Eq.1 just correlated the dynamic viscosity to temperature without structural information. In this study, attention was paid on the Martin's rule of free energy additivity. The free energy of viscous flow for n-Alkane-1-ol in Eq.1 was expanded to the enthalpy and entropy forms.

## 2. Theory

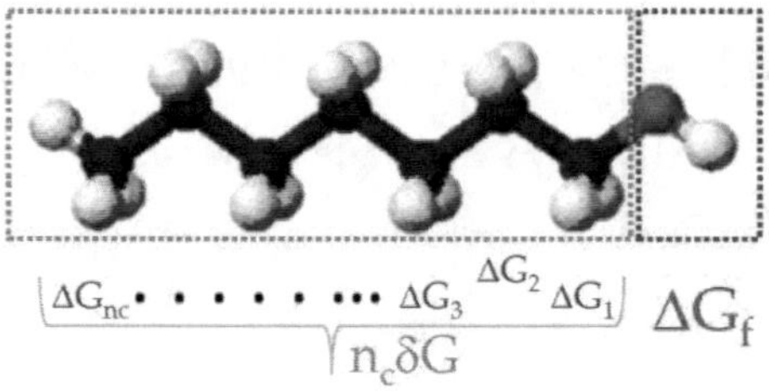


Figure 1 correlation of free energy additivity and n-alkan-1-ol molecular structure.

From figure 1 show a compound having the molecular structure of $CH_3(-CH_2)_{nc}$-1-*f*. Martin [4] divided the molecule into different groups: *f*, $CH_2$, $CH_3$. The free energy of transfer from solution to gas ($\Delta G$, in gas chromatography) of the molecule was derived from the sum of the free energies of all contributed groups; ie, the free energy contribution from the structural groups are additive.

$$\Delta G = \Delta G_f + \Delta G_1 + \Delta G_2 ...... + \Delta G_{n_c} \tag{3}$$

$\Delta G_1$.....$\Delta G_{nc}$ are the free energies of the methylene and methyl groups, which are not very different. Thus, they are average to $\Delta G$ and Eq.3 is shortening to Eq.4.

$$\Delta G = \Delta G_f + n_c \delta G \tag{4}$$

where $\Delta G_f$ is the free energy of the functional group, *f*; $\Delta G$ is the change in free energy/carbon atom; $n_c$ is the number of carbon atoms.

Substitution Eq.4 into Eq.1and expansion the free energy to enthalpy and entropy forms (G=H-TS),

$$\ln\eta = \ln A - \frac{\Delta H_f}{RT} + \frac{\Delta S_f}{R} - \frac{n_c \delta H}{RT} + \frac{n_c \delta S}{R} \tag{5}$$

or

$$\ln\eta = a + bn_c + (c + dn_c)\frac{1}{T} \tag{6}$$

where $a = \ln A + \frac{\Delta S_f}{R}$, $b = \frac{\delta S}{R}$, $c = -\frac{\Delta H_f}{R}$ and $d = -\frac{\delta H}{R}$

Eq.6 is used for estimation of dynamic viscosity of n-alkan-1-ol at several temperatures.

## 3. Materials and Methods

### *3.1 Dynamic viscosity data*

The dynamic viscosities values of n-alkan-1-ol were obtained from the report of Mokhtaram et al.[5] (propanol and butanol) and Baltazer et al.[6]. (pentanol, hexanol and heptanol). The dynamic viscosities were measured by using Anton Paar Automated Micro Viscometer (AMVn).

### *3.2 Numeric constants of Eq.6*

All the four numeric constants of Eq.5 were solved by multiple linear regression according to Phankosol et al. [3, 7, 8] as shown in Eq.7.

$$\ln\eta = -5.008 + 1481.8n_C - (0.3925 + 194.07n_C)\frac{1}{T} \qquad (7)$$

### *3.3 Statistical analysis*

The percent average absolute deviations (AAD (%)) were calculated from Eq.8.

$$AAD(\%) = \frac{1}{N}\sum_{i=1}^{N}\left[\frac{|\eta_{exp} - \eta_{cal}|}{\eta_{exp}} x100\right] \qquad (8)$$

where *exp* stands for experimental value reported elsewhere, *cal* is the calculated value and *N* is the number of data points.

## 4. Results and Discussion

Percent differences between the calculated by Eq.7 and literature dynamic viscosities values are listed in the parentheses, at 288.15-363.15 K are summarized in Table 1. The estimated dynamic viscosities values for n-propanol, n-butanol, n-pentanol, n-hexanol and n-heptanol agrees well with the literature values. The highest absolute difference is 6.65% (Propanol at 333.15 K). The Bias and AAD (%) were -0.17% and 1.73%, respectively.

Table 1 Estimated dynamic viscosities (mPa.s) of n-alkan-1-ol at 288.15−363.15 K

| T (K) | Propanol[5] ($n_c$=3) | | 1-Butanol[5] ($n_c$=4) | | 1-Pentanol[6] ($n_c$=5) | | 1-Hexanol[6] ($n_c$=6) | | 1-Heptanol[6] ($n_c$=7) | |
|---|---|---|---|---|---|---|---|---|---|---|
| | $\eta_{cal.}$ (mPa.s) | D (%) | $\eta_{cal.}$ (mPa.s) | D (%) | $\eta_{cal.}$ (mPa.s) | D (%) | $\eta_{cal.}$ (mPa.s) | D (%) | $\eta_{cal.}$ (mPa.s) | D (%) |
| 288.15 | 3.52 | 3.97 | 3.52 | -3.80 | 4.66 | - | 6.17 | - | 8.18 | - |
| 293.15 | 2.35 | 2.42 | 3.08 | -2.31 | 4.03 | -0.01 | 5.28 | 1.54 | 6.91 | 2.91 |
| 298.15 | 2.09 | 1.31 | 2.71 | -3.85 | 3.50 | -0.59 | 4.54 | 0.85 | 5.88 | 2.02 |
| 303.15 | 1.86 | 0.51 | 2.39 | -3.19 | 3.06 | -1.12 | 3.92 | 0.31 | 5.02 | 1.34 |
| 308.15 | 1.67 | 0.08 | 2.12 | -2.48 | 2.68 | -1.65 | 3.40 | -0.15 | 4.31 | 0.82 |
| 313.15 | 1.50 | 1.54 | 1.88 | -1.92 | 2.36 | -2.13 | 2.97 | -0.55 | 3.72 | 0.45 |
| 318.15 | 1.35 | 2.22 | 1.68 | -1.30 | 2.09 | -2.59 | 2.60 | -0.87 | 3.23 | 0.23 |
| 323.15 | 1.22 | 2.59 | 1.51 | -0.75 | 1.85 | -2.99 | 2.28 | -1.10 | 2.81 | 0.11 |
| 328.15 | 1.11 | 3.67 | 1.35 | 0.18 | 1.65 | -3.40 | 2.02 | -1.25 | 2.46 | 0.14 |
| 333.15 | 1.01 | 6.65 | 1.22 | 1.78 | 1.48 | -3.79 | 1.79 | -1.31 | 2.16 | 0.24 |
| 338.15 | 0.92 | - | 1.10 | - | 1.32 | -4.14 | 1.59 | -1.35 | 1.90 | 0.47 |
| 343.15 | 0.84 | - | 1.00 | - | 1.19 | -4.46 | 1.42 | -1.29 | 1.68 | 0.84 |
| 348.15 | 0.77 | - | 0.91 | - | 1.08 | - | 1.27 | -1.15 | 1.50 | 1.26 |
| 353.15 | 0.71 | - | 0.83 | - | 0.97 | - | 1.14 | -0.98 | 1.33 | 1.79 |
| 358.15 | 0.66 | - | 0.76 | - | 0.88 | - | 1.03 | -0.78 | 1.19 | 2.36 |
| 363.15 | 0.61 | - | 0.70 | - | 0.80 | - | 0.93 | -0.51 | 1.07 | 3.00 |
| **Overall** | **Bias** | **2.50** | | **-1.76** | | **-2.44** | | **-0.57** | | **1.20** |
| | **AD (%)** | **2.50** | | **2.16** | | **2.44** | | **0.93** | | **1.20** |

The correlation between the reported experimental dynamic viscosities of Mokhtaram et al.[5] and Baltazer et al.[6] and estimated values using Eq.7 (61 data points) show in figure 2 was linear with the slope of 0.982 and intercept at 0.0427. The $R^2$ and standard error are 0.999 and 0.022, respectively.

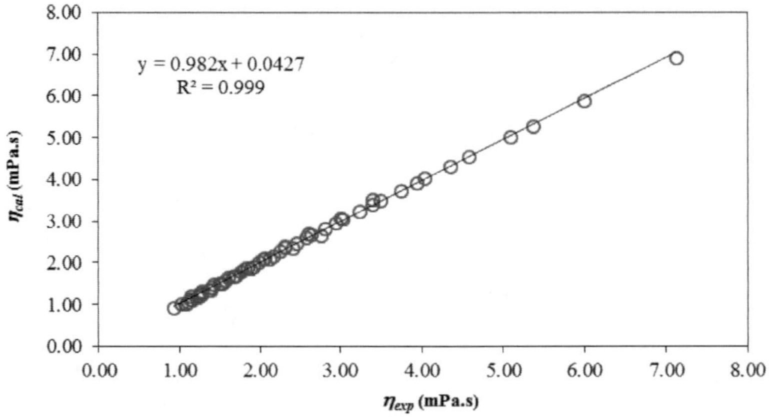


Figure 2 Correlation of the estimated dynamic viscosities ($\eta_{cal}$) to the experiment values ($\eta_{exp}$) [3,10] of n-alkan-1-ol at 288.15-363.15 K.

## 5. Conclusions

This work, the total free energy is then related to physical property of the n-alkan-1-ol via an existing equation. Thus, the derive equation can then be used to estimate dynamic viscosities of the n-alkan-1-ol at different temperatures with good accuracy. The dynamics viscosities values outside of this temperature range may be possibly predicted by the correlation but the accuracy may be lower.

**Acknowledgements**

This work was supported by Research and Development Institute Bansomdejchopraya Rajabhat University.

**References**

[1] Alaoui F E M, Montero E A, Qiu G, Aguilar F and Wu J 2013 Liquid density of biofuel mixtures: 1-Heptanol + heptane system at pressures up to 140 MPa and temperatures from 298.15 K to 393.15 K *The Journal of Chemical Thermodynamics* **65** 174-83

[2] Krisnangkura K, Aryusuk K, Phankosol S and Lilitchan S 2016 Energy Additivity Approaches to QSPR Modeling in Estimation of Dynamic Viscosity of Fatty Acid Methyl Ester and Biodiesel *Journal of the American Oil Chemists' Society* **93** 1407-14

[3] Phankosol S, Sudaprasert K, Lilitchan S, Aryusuk K and Krisnangkura K 2015 An Empirical Equation for Estimation of Kinematic Viscosity of Fatty Acid Methyl Esters and Biodiesel *Journal of the American Oil Chemists' Society* **92** 1051-61

[4] Martin A J P 1950 Partition Chromatography *Annual Review of Biochemistry* **19** 517-42

[5] Mokhtarani B, Sharifi A, Mortaheb H R, Mirzaei M, Mafi M and Sadeghian F 2009 Density and viscosity of 1-butyl-3-methylimidazolium nitrate with ethanol, 1-propanol, or 1-butanol at several temperatures *The Journal of Chemical Thermodynamics* **41** 1432-8

[6] Estrada-Baltazar A, Bravo-Sanchez M G, Iglesias-Silva G A, Alvarado J F J, Castrejon-Gonzalez E O and Ramos-Estrada M 2015 Densities and viscosities of binary mixtures of n-decane + 1-pentanol, + 1-hexanol, + 1-heptanol at temperatures from 293.15 to 363.15 K and atmospheric pressure *Chinese Journal of Chemical Engineering* **23** 559-71

[7] Phankosol S, Sudaprasert K, Lilitchan S, Aryusuk K and Krisnangkura K 2014 Estimation of surface tension of fatty acid methyl ester and biodiesel at different temperatures *Fuel* **126** 162-8

[8] Phankosol S, Sudaprasert K, Lilitchan S, Aryusuk K and Krisnangkura K 2014 Estimation of Density of Biodiesel *Energy & Fuels* **28** 4633-41

# Microanalysis study on ancient Wiangkalong Pottery

**K Won-in[1], S Tancharakorn[2] and P Dararutana[3,*]**

[1] Department of Earth Sciences, Faculty of Science, Kasertsart University, Bangkok 10900 Thailand
[2]Synchrotron Light Research Institute, Muang District, Nakhon Ratchasima 30000 Thailand
[3]Royal Thai Army Chemical Department, Bangkok 10900 Thailan
[*]E-mail: pisutti@hotmail.com

**Abstract.** Wiangkalong is one of major ceramic production cities in northern of Thailand, once colonized by the ancient Lanna Kingdom (1290 A.D.). Ancient Wiangkalong potteries were produced with shapes and designs as similar as those of the Chinese Yuan and Ming Dynasties. Due to the complex nature of materials and objects, extremely sensitive, spatially resolved, multi-elemental and versatile analytical instruments using non-destructive and non-sampling methods to analyze theirs composition are need. In this work, micro-beam X-ray fluorescence spectroscopy based on synchrotron radiation was firstly used to characterize the elemental composition of the ancient Wiangkalong pottery. The results showed the variations in elemental composition of the body matrix, the glaze and the underglaze painting, such as K, Ca, Ti, V, Cr, Mn and Fe.

## 1. Introduction

Ceramics have been made from naturally using raw materials for centuries in Thailand. In ancient time, it was known that clay minerals became plastic when water was added and could be molded in different shapes. After drying with the sun then they fired at high temperature mainly using fire wood.

The study of ancient ceramics, as a branch of archaeology, has centered on establishing the data and area of production of pottery artifacts. Typological classifications, technological observations and comparative studies serve to clarify the development and cultural inter-relationships of various classes of pottery objects.

In Lanna region, there are more than ten known ancient ceramic kiln sites. Wiangkalong is one of them which located at Chiang Rai Province and is dated to the $15^{th}$-$16^{th}$ century A.D. There are four categories of glazed ceramic products manufactured such as underglaze painted black-on-white ware, white ware with colorless glaze, green-glaze ware and brown ware. Ceramic ware of the Wiangkalong kiln has a fine stoneware white, buff or gray body. Overall glazing is applied to the inner and outer walls down to the edge of the foot-rim leaving the exterior base underglaze and bearing fine striation marks. Underglaze painted black-on-white wares are major and unique group of Wiangkalong ware, manufactured throughout the district and similar to the designs on Chinese cobalt-decorated porcelains of the Ming Dynasty. White wares with transparent colorless glaze or monochrome-white constitute a second category and are characterized by a thin and fragile stoneware body with a transparent colorless glaze similar to white ware of the Sung Dynasty. Green-glaze wares or Wiangkalong celadon are a relatively rare category which were similar in shape and color of the glaze to celadon ware of the Yuan and Ming Dynasties. Brown wares are rare and have no identity or special characteristics. Glaze color varies from reddish brown to yellowish brown, as shown in **Figure 1** [1].

IOP Conf. Series: Journal of Physics: Conf. Series **901** (2017) 012024 doi:10.1088/1742-6596/901/1/012024

**Figure 1.** Types of Wiangkalong pottery

Previous study of Wiangkalong pottery using Raman spectroscopy showed that the body matrix mineralogy was highly heterogeneous, suggesting the presence of large variety of minerals and trace elements in the clay, as well as the variable conditions of firing control [2].

## 2. Experimental

Potsherd that found at Wiangkalong which was a underglaze painted black-on-white ware that mainly excavated from the Wiangkalong site (**Figure 2**), was characterized by using μ-XRF. A gray- body was decorated with monochrome underglaze in black color and coated with transparent colorless glaze. Three areas were selected, as shown in **Figure 3**. They were marked as position 1, 2 and 3 for the glaze, body and underglaze painting, respectively.

**Figure 2.** Wiangkalong site

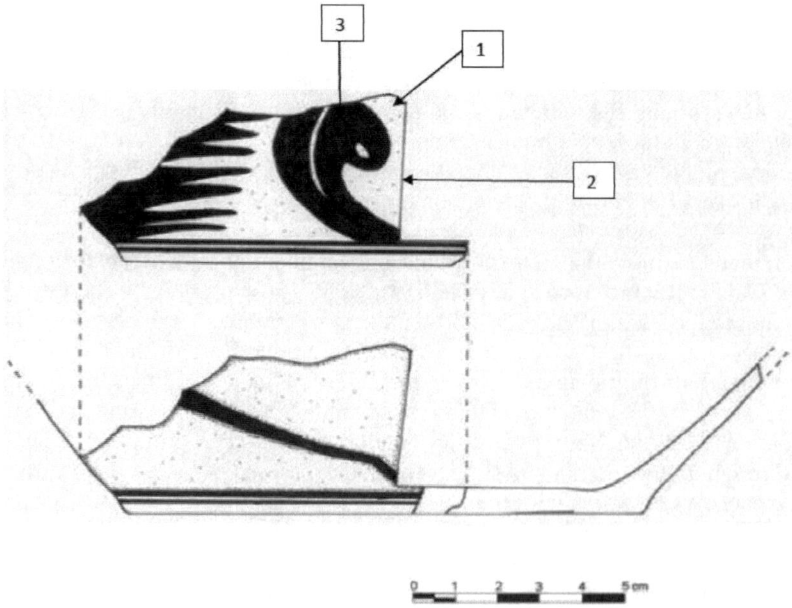


**Figure 3.** Potsherd of Wiangkalong underglaze painted black-on-white ware

Micro-beam energy dispersive X-ray fluorescence spectroscopy (μ-XRF) using synchrotron radiation at BL6b of the Synchrotron Light Research Institute, Thailand was also used to confirm the presence of metals in the sample. The BL6b utilizes continuous synchrotron radiation from the bending magnet with energy range above 2 keV with the beam size of 100 μm at the sample position using polycapillary half-lens as an X-ray optics. To perform the experiment, the sample is placed on a three-degree of freedom motorized stage. Using high precision motorized stage allows us to accurately specify area or point of interest on the sample.

## 3. Results and Discussion

Results from micro-XRF provided qualitative information about the elemental composition of the ancient grain. However, due to the limitation of our XRF system, elements with low atomic numbers (Z<19) such as carbon (C), aluminum (Al) and silicon (Si) were not be detected. The spectra of the potsherd of Wiangkalong underglaze painted black-on-white ware in **Table 1** and **Figure 4** showed the existence of some metals including potassium (K), calcium (Ca), titanium (Ti), vanadium (V), chromium (Cr), manganese (Mn) and iron (Fe).

**Table 1.** Elemental composition of potsherd of Wiangkalong underglaze painted black-on-white ware using μ-XRF

| Position | Detectable elements |
|---|---|
| Glaze | K, Ca, Ti, V, Cr, Mn, Fe |
| Body | K, Ca, Ti, V, Cr, Mn, Fe |
| Underglaze painting | K, Ca, Ti, V, Cr, Mn, Fe |

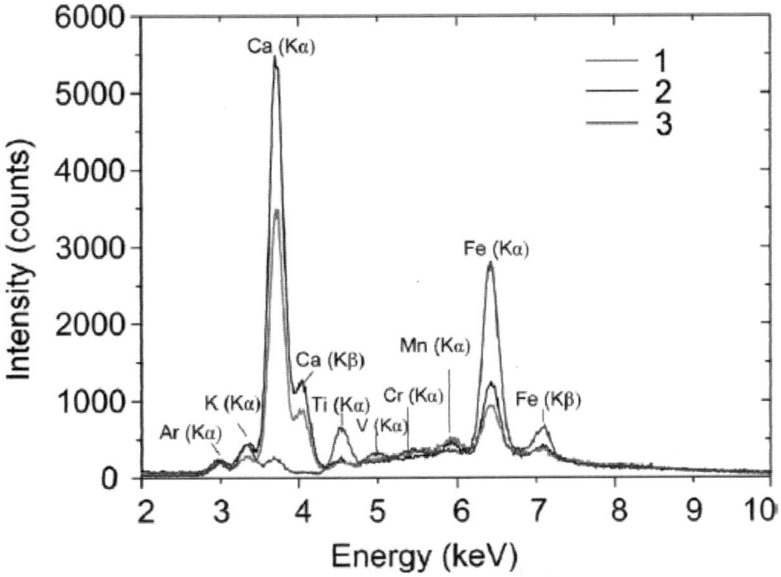


**1= Glaze, 2= Body, 3= Underglaze painting**

**Figure 4.** XRF spectra of potsherd of Wiangkalong underglaze painted black-on-white ware

It is also clearly seen that high concentration of iron has been found at the underglaze painting (Position 3). It indicates that iron is one the chemical composition in black colorant. Noted that Ar(Ka) peaks shown in **Figure 4** come from the sample environment.

It is planned to do some advance studies and more samples in order to analyze chemical composition and elemental distribution using other techniques, such as scanning electron microscope coupled with energy dispersive X-ray fluorescence spectroscopy (SEM-EDS), proton-induced X-ray emission spectroscopy (PIXE), X-ray diffraction (XRD) Infrared spectroscopy (IR).

## 4. Conclusions

Micro-beam XRF can be effectively used to identify elemental composition in a specific area on a sample. Most of the medium atomic numbers which are the main components of the potsherd of Wiangkalong underglaze painted black-on-white ware can be detected even low concentration such as titanium (Ti), vanadium (V) and chromium (Cr). These elements may be used as the fingerprint elements.

This non-destructive technique provided a simple and quick method for qualitative elemental analysis. This was the first time that this technique was used to study and obtained complementary information about the ancient pottery in Thailand.

## 5. Acknowledgments

This work was partly funded by the Faculty of Science at Kasetsart University and the Office of National Research Council of Thailand. Authors were kindly thanks the Faculty of Archaeology at Silpakorn University for supporting the pottery sample. The Synchrotron Light Research Institute at Nakhon Ratchasima was also thanked for providing μ-XRF.

## References

[1] S. Praichanjit. 2011. Archaeology of Ceramics in Lanna Northern Siam. (Bangkok: Samaphun Publishing Co.Ltd.)

[2] K. Won-in, S. Pongkrapan, P. Dararutana, M. Thawwornmongkolkij and P. Wathanakul. 2010. Raman spectroscopic study of ancient potteries in Thailand: Wiangkalong pottery, 5$^{th}$ Mid-European Clay Conference; August 25-29, 2010: Budapest, Hungary.

# Unpinning of spiral waves from rectangular obstacles by stimulated wave trains

**Benjamas Ponboonjaroenchai[1], Panatda Srithamma[1], Nakorn Kumchaiseemak[1], Malee Sutthiopad[1], Stefan C. Müller[2], Chaiya Luengviriya[1] and Jiraporn Luengviriya[3,*]**

[1] Department of Physics, Kasetsart University, 50 Phaholyothin Road, Jatujak, Bangkok 10900, Thailand
[2] Institute of Experimental Physics, Otto-von-Guericke University Magdeburg, Universitätsplatz 2, D-39106 Magdeburg, Germany
[3] Lasers and Optics Research Center, Department of Industrial Physics and Medical Instrumentation, King Mongkut's University of Technology North Bangkok, 1518 Pibulsongkram Road, Bangkok 10800, Thailand

* E-mail: jiraporn.l@sci.kmutnb.ac.th

**Abstract** Pinned spiral waves are exhibited in many excitable media. In cardiology, lengthened tachycardia correspond to propagating action potential in forms of spiral waves pinned to anatomical obstacles including veins and scares. Thus, elimination such waves is important particularly in medical treatments. We present study of unpinning of a spiral wave by a wave train initiated by periodic stimuli at a given location. The spiral wave is forced to leave the rectangular obstacle when the period of the wave train is shorter than a threshold $T_{unpin}$. For small obstacles, $T_{unpin}$ decreases when the obstacle size is increased. Furthermore, $T_{unpin}$ depends on the obstacle orientation with respect to the wave train propagation. For large obstacles, $T_{unpin}$ is independent to the obstacle size. It implies that the orientation of the obstacle plays an important role in the unpinning of the spiral wave, especially for small rectangular obstacles.

## 1. Introduction

Spiral waves have been observed in various excitable media [1-5]. In the heart tissues, the spiral waves are the origin of ventricular tachycardia and fibrillation which potentially leading to sudden cardiac death [6]. They will survive much longer if they are pinned to anatomical inhomogeneities or obstacles, e.g., dead cells or scars [1]. Cardioversion shocks for eliminating spiral waves from the heart are applications of high electrical voltage. The method results in damage of the cardiac tissues.

To avoid this problem, many studies have been performed to eliminate the spiral waves with low voltage, e.g., applications of wave trains. It is found that high-frequency wave trains induce drift of free spiral waves [7] and they can unpin spiral waves from circular obstacles [8,9]. Such drifting spiral waves are annihilated when the spiral tip hit the boundary.

In this article, we present an elimination of spiral waves pinned to rectangular obstacles with different orientations with respect to the direction of the applied wave trains in simulations by using the Oregonator model [10].

## 2. Methods

In our simulations, the Oregonator model is used to describe the dynamics of the activator u and the inhibitor v in excitable media as shown in Eq. 1

$$\frac{\partial u}{\partial t} = \frac{1}{\varepsilon}\left(u - u^2 - fv\frac{u-q}{u+q}\right) + D_u\nabla^2 u,$$

$$\frac{\partial v}{\partial t} = u - v + D_v\nabla^2 v. \quad (1)$$

As in a study by Jahnke and Winfree [11], the parameters were chosen as ε = 0.01, q = 0.002, f = 1.4, and the diffusion coefficients as $D_u$ = 1.0 and $D_v$ = 0.6. For this parameter set, the system supported spiral waves with a circular spiral core of 0.9 space units (s.u.) in diameter, wavelength = 10.5 s.u., period = 1.55 time units (t.u.), and velocity = 6.76 s.u. t.u.$^{-1}$.

The variables u and v in Eq. 1 were calculated using an explicit Euler method with a 9-point approximation of the two-dimensional Laplacian operator on a discrete system of a dimensionless size = 40×40 s.u. with a uniform grid space of Δx = Δy = 0.1 s.u. and a time step Δt = 3.0 x $10^{-3}$ t.u., as required for numerical stability ($\Delta t \leq (3/8)(\Delta x)^2$ [12]). A single unexcitable rectangle was defined as the obstacle in each simulation. The boundaries of both the medium and the obstacle had no-flux conditions.

We tested totally twenty rectangles consist of two series: The first series (fixed height) 0.5×2.0, 1.0×2.0, 1.5×2.0, 2.0×2.0, 2.5×2.0, 3.0×2.0, 3.5×2.0, 4.0×2.0, 4.5×2.0 and 5.0×2.0 s.u.$^2$ and the second series (fixed width) 2.0×0.5, 2.0×1.0, 2.0×1.5, 2.0×2.0, 2.0×2.5, 2.0×3.0, 2.0×3.5, 2.0×4.0, 2.0×4.5 and 2.0×5.0 s.u.$^2$.

To create a spiral wave, a planar wave was triggered by setting a 5-grid-point strip at a boundary to an excited state (u = 1.0 and v = 0.0) and subsequently reset a half of it to an excitable state (u = 0.0 and v = 0.0) when it reaches the obstacle at the middle of the medium. A periodic wave train is simulated by setting a semicircular area (4.0 s.u. in diameter) to an excited state at the left edge of the medium.

## 3. Results

Figure 1 shows three examples of spiral waves eliminated by wave trains. The top row [figures 1(a)-1(d)] illustrates the phenomenon in the absence of obstacles. At the beginning, the tip of the free spiral wave rotates around a circular core (diameter 0.9 s.u.) at the middle of the medium [figure 1(a)]. When the fronts of the stimulated wave train with a period = 1.51 t.u. (slightly shorter than the period $T_{sp}$ of the free spiral wave 1.55 t.u.) reach the tip of free spiral wave [figure 1(b)], the position of the spiral tip drifts to the right of the medium [see the red line in figure 1(c)] and then the spiral wave is terminated at the boundary [figure 1(d)].

In the middle row [figure 1(e)-1(h)], the spiral wave is initially pinned to a rectangular obstacle with 2.0 s.u. × 0.5 s.u. in dimensions [figure 1(e)]. This pinned spiral wave has a period $T_{sp}$ of 1.59 t.u. which longer than that of the free spiral wave, as reported earlier [13]. To unpin the spiral wave i.e. its tip is detached from the obstacle [figure 1(g)], the period of the wave train must be lower than a critical value called $T_{unpin}$ of 1.48 t.u. which is shorter than that in the case of the free spiral wave (the top row). Eventually, the spiral wave is forced to be annihilated at the boundary. Note that this pinned spiral wave is harder to be eliminated in comparison to the free spiral wave since the wave train is required to have a shorter period.

In the bottom row [figure 1(i)-1(l)], the spiral wave is pinned to a rectangular obstacle with 0.5 s.u. × 2.0 s.u. in dimensions [figure 1(e)]. This obstacle has the same size but different orientation to that in the middle row so that these two pinned spiral waves has the same wave period. Surprisingly, the unpinning of spiral wave occurs at a period of the wave train $T_{unpin}$ = 1.40 t.u., i.e., shorter than that in the middle row.

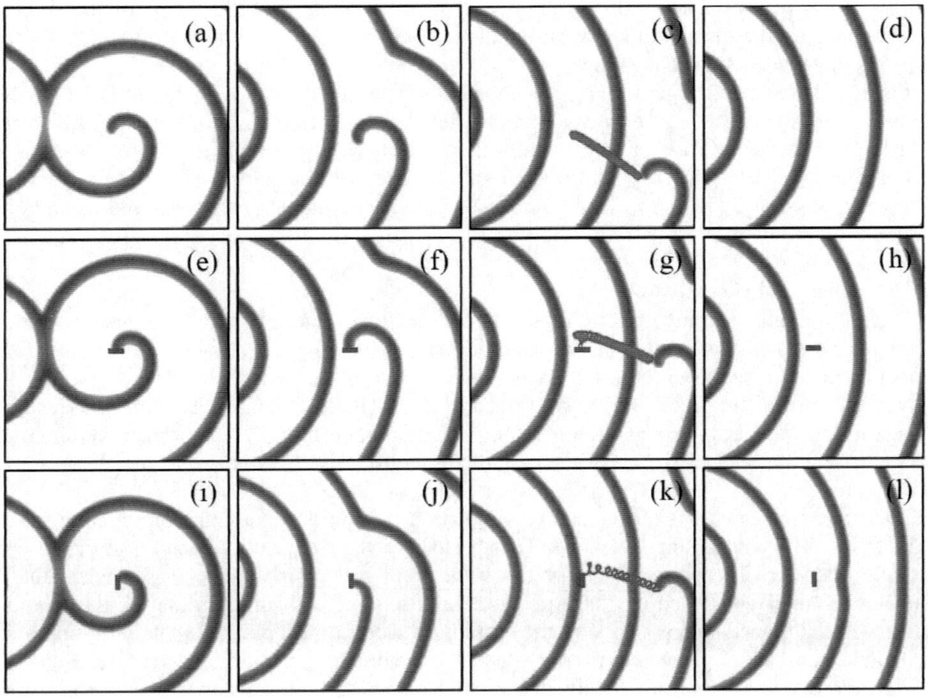


**Figure 1.** Elimination of spiral waves by a wave train application: (a)-(d) A free spiral wave is forced to drift and be annihilated by a wave train with a period of 1.51 t.u., Pinned spiral wave are forced to unpin from rectangles with dimensions (e)-(h) 2.0 s.u. × 0.5 s.u. and (i)-(l) 0.5 s.u. × 2.0 s.u. and subsequently to be annihilated by wave trains with different periods of 1.48 t.u. and 1.40 t.u. The red paths indicate the trajectories of the spiral tips.

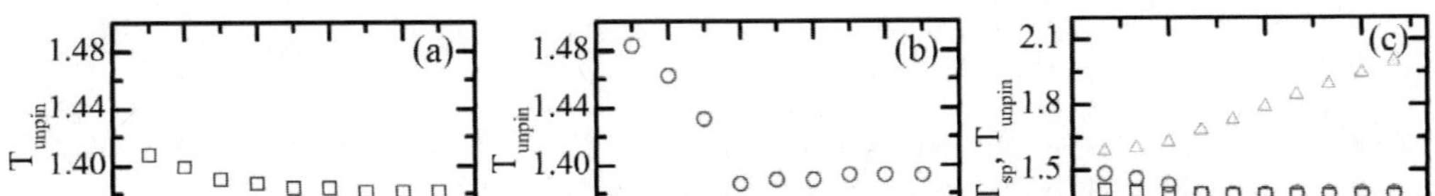


**Figure 2**. The critical period $T_{unpin}$ of the wave train for unpinning of spiral wave vs the obstacle (a) width x, (b) height y, and (c) area A with squares and circles indicating $T_{unpin}$ taken from (a) and (b), respectively. Triangles in (c) represent $T_{sp}$ the period of the spiral waves pinned to different obstacles.

The detailed results of the unpinning of spiral waves by wave trains in our simulations are presented in figure 2. For the first series of obstacles with a fixed height of 2.0 s.u., the critical value of wave train for unpinning the spiral wave $T_{unpin}$ decreased slightly while the obstacle width X is increased [figure 2(a)]. The simulations with the second series of obstacles with a fixed width of 2.0 s.u. give different results [figure 2(b)]. As the obstacle height Y is increased, $T_{unpin}$ decreased very much for Y < 2.0 s.u. but $T_{unpin}$ is approximately constant for Y > 2.0 s.u.

To compare results from all obstacles, $T_{unpin}$ is plotted against the obstacle area A [circles and squares in figure 2(c)]. Clearly, in these unpinning phenomena the obstacle height is more crucial in comparison to the obstacle width especially for small rectangle obstacles. This implies that for such small rectangles, the success of spiral elimination depends on also the orientation of the obstacles with respect to the wave train propagation.

Finally, the plot of the period $T_{sp}$ of pinned spiral waves [triangles in figure 2(c)] shows that it increases with the obstacle size as described earlier in [13]. Termination of these pinned spiral waves is distinguish from the case of free spiral waves which can be illuminated by a wave train with a period slightly shorter than $T_{sp}$ of the free spiral waves. When the obstacle size is increased, the difference between $T_{sp}$ and $T_{unpin}$ also increases [compare triangles to circles and squares in figure 2(c)].

## 4. Discussion and Conclusion

We have presented an investigation on the termination of spiral waves pinned to unexcitable rectangular obstacles with different sizes and orientations by an application of wave trains. The results show that the elimination of the pinned spiral wave requires the period of the wave train $T_{unpin}$ shorter than in the case of the free spiral wave. For a given small rectangular obstacle, its orientation with respect to the wave train propagation also plays an important role, i.e., the longer rectangle facing to the wave train, the shorter $T_{unpin}$. For the case of large obstacles, $T_{unpin}$ is almost independent to the size of the rectangles.

Recently, Tanaka et al. [8,9] presented simulations of spiral unpinning from circular obstacles by wave trains. The pinned spiral waves have a period, wavelength, and velocity that increase with the obstacle diameter. The period $T_{unpin}$ of the wave train for unpinning decreases when the obstacle diameter is increased. Based on these results, they proposed a theory to describe the relation of the velocity of the pinned spiral wave and the required period $T_{unpin}$. The results of spiral unpinning from rectangular obstacles, presented in our study, are more complicated so that we suggest further investigations on this topic to obtain deeper understanding which leads to a generalization of the unpinning theory.

**Acknowledgments**

We thank the Faculty of Science, the Research and Development Institute (KURDI), the Center for Advanced Studies of Industrial Technology, and the Graduate School, Kasetsart University; and the Office of the Higher Education Commission and King Mongkut's University of Technology North Bangkok (contract no. KMUTNB- NRU-58-03) for financial support.

## References

[1] Davidenko J M, Pertsov A M, Salomonz R, Baxter W and Jalife J 1992 *Nature* **335** 349
[2] Dahlem M A and Müller S C 1997 *Exp. Brain Res* **115** 319-324
[3] Siegert F and Weijer C J 1989 *Cell Sci* **93** 325-335
[4] Winfree A T 1972 *Science* **175** 634-636
[5] Müller S C, Plesser T and Hess B 1985 *Science* **230** 661-663
[6] Cherry E M and Fenton F H 2008 *New J. Phys* **10** 125016
[7] Dutta S and Steinbock O 2011 *Phys. Rev. E* **83** 0562131-8
[8] Tanaka M, Isomura A, Hörning M, Kitahata H and Agladze K 2009 *Chaos* **19** 043111-5
[9] Tanaka M, Hörning M, Kitahata H and Yoshikawa K 2015 *Chaos* **25** 1031271-9
[10] Field R J and Noyes R M 1974 *J. Chem. Phys.* **60** 1877-1884
[11] Jahnke W, Skaggs W E and Winfree A T 1989 *J Phys Chem* **93** 740-749
[12] Dowle M, Mantel R M and Barkley D 1997 *Chaos* **7** 2529
[13] Sutthiopad M, Luengviriya J, Porjai P, Phantu M, Kanchanawarin J, Müller S C and Luengviriya C 2015 *Phys Rev E* **91** 052912

# Preparation of strontium- and/or zinc-doped hydroxyapatite nanoparticles and their polycaprolactone composite fibrous scaffolds

**Teeranat Pantasri[1], Suphasinee Seet[2], and Pimsiree Suwanna[2,3*]**

[1] Materials Science and Nanotechnology Program, Faculty of Science, Khon Kaen University, Khon Kaen, Thailand
[2] Integrated Nanotechnology Research Center, Khon Kaen University, Khon Kaen, Thailand
[3] Department of Materials Science, Faculty of Science, Kasetsart University, Bangkok, Thailand

[*]E-mail: pimsiree.s@ku.ac.th

**Abstract**. In this work, hydroxyapatite (HA) and hydroxyapatite doped with Sr (HA-Sr), Zn (HA-Zn) and both Sr-Zn (HA-SrZn) were synthesized by a sol-gel method and combined with polycaprolactone (PCL) to make HA/PCL composites using an electrospinning technique. The synthesized nanoparticles and their composite fibers were investigated using various techniques. The X-Ray Diffraction (XRD) result showed the characteristic peaks of the hydroxyapatite structure; whereas the scanning electron microscopy (SEM) and transmission electron microscopy (TEM) results revealed that the synthesized nanoparticles were successfully incorporated into the randomly interconnected and highly porous PCL matrix.

## 1. Introduction

An ideal scaffold for bone regeneration should mimic both the structure and mechanical properties of the natural bone. It should also be bioresorbable with a controllable degradation rate to match that of the regeneration rate of the newly-formed bone tissue [1]. Hydroxyapatite (HA) has received high attention due to its chemical similarity to the inorganic matrix of natural bone, excellent osteoconductivity and bioactivity. However, pure HA have low bioactive property due to their low resorbability [2]. Many studies on modifying HA composition have been conducted to improve its properties such as bioactivity, degradation behavior and antibacterial property [3-6].

In this work, strontium (Sr) and zinc (Zn) were chosen to be doped into the structure of HA due to their noble characteristics. Sr plays an important role in the treatment of osteoporosis and enhancement of bone remineralization as it is associated with a reduction of bone resorption and an increase of new bone formation [5] while Zn plays a vital role in the biochemistry of bone tissues and exhibits antibacterial activity [6]. Poly(ε-caprolactone) (PCL) was chosen to be a matrix for composite fiber since it is a biocompatible and has been widely used in many biomedical applications [7].

## 2. Materials and Methods

### *2.1. Materials*

Calcium nitrate tetrahydrate (CNT, 99.0% purity), Diammonium hydrogen phosphate (DAHP, 99.0% purity) were purchased from Qrec, Malaysia. Strontium acetate (SA, 99.9% purity), Zinc acetate (ZA, 99.9% purity) and Poly(ε-caprolactone) (PCL, MW=80,000) were obtained from Sigma-Aldrich, USA. Dichloromethane (DCM, 99.9% purity) and N,N-dimethylformamide (DMF, 99.8% purity) were purchased from Showa, Japan. Ethanol (95.0% purity) was supplied by Merck.

*2.2 Preparation and general characterization of HA, HA-Sr, HA-Zn and HA-SrZn particles*
HA, HA-Sr, HA-Zn, and HA-SrZn powders were synthesized using a sol-gel method [8]. The precursors were mixed at the (Ca+Sr/Zn):P mole ratio of 10:6 and at the Ca:Sr/Zn ratio of 9:1. In the case of HA-ZnSr, the Sr:Zn mole ratio was 1:1. CNT was dissolved in ethanol, and where applicable SA and/or ZA were added. DAHP was separately dissolved in deionized water. The two solutions were quickly mixed thoroughly after which the gel was formed. The gel was then aged for 24 hours at room temperature and heated at 80 °C. The obtained powder was then calcined at 600 °C for 2 hours.

The X-ray diffractometer (XRD, D8 ADVANCE, Bruker) was used to identify the phase structure of the calcined powders. The XRD patterns were obtained in a 2θ angle ranging from 20° to 80° at a scanning speed of 0.02°/min. To observe their morphologies, the calcined powders were dispersed in ethanol by sonication for 15 min, dropped onto the carbon coated copper grids and air-dried for a transmission electron microscopy (TEM, TECNAI G220, FEI) observation.

*2.3 Electrospinning and general characterization of PCL/doped and undoped-hydroxyapatite composite fibrous scaffolds*
A 10% w/v PCL solution was prepared by dissolving PCL in a DCM/DMF mixture (DCM:DMF volume ratio of 20:80). The HA, HA-Sr, HA-Zn or HA-SrZn powder was then added into the PCL solution at the HA:PCL weight ratio of 1:10 and mixed well for 3 hours using an ultrasonic bath.

The PCL/HA composite fibrous scaffold was prepared by electrospinning (KKU ElectroSys) of the mixture suspension. The as-prepared suspension was added into a 10 ml plastic syringe mounted onto a syringe pump in which the needle was connected to a high-voltage supply. Under 18 kV voltage, the fluid jet was injected out at a rate of 1.0 ml/h and the resultant composite fibers were collected on an aluminum foil at the distance of 20 cm from the needle. The morphologies of the fibrous scaffolds were studied using a scanning electron microscope (SEM, S-3000N, Hitachi) and a TEM. For the SEM observation, each scaffold was sputter-coated with a thin layer of gold. For the TEM observation, the fibers were electrospun directly onto the carbon coated copper grids and air-dried.

## 3. Results and Discussion

*3.1 Phase and morphology of the doped and undoped hydroxyapatite particles*
Figure 1 shows the XRD patterns of the synthesized powders. Hydroxyapatite (JCPDs 74-0566) was found to be the main phase for all compositions with β-tricalcium phosphate (JCPDs 09-0169) as a second phase. In addition, α-tricalcium phosphate (JCPDs 09-0348) and $CaZn_2(PO_4)_2(H_2O)_2$ (JCPDS 86-2372) were also observed in the HA-Zn and HA-SrZn powders.

The morphology of synthesized particles can be seen from the TEM images in Figure 2. In general, the undoped HA particles are relatively equiaxed. The particle sizes are approximately in the range of 80-280 nm, with the average of 151±47 nm. More variety can be observed in cases of doped HA samples both in terms of particle shape and size. In addition to the equiaxed grains, rod-like particles are also evident, most prominently in the case of HA-SrZn. The size of the equiaxed particles appear to be smaller than those of the pure HA, as observed to be in the range of 50-200 nm and with the averages of 116±31, 86±45, and 94±31 nm in diameter for the HA-Sr, HA-Zn and HA-SrZn samples, respectively. The width and the length of the rod-like particles are approximately 50-200 nm and 150-300 nm, respectively, and an aspect ratio is approximately 1.8. This is consistent with the work of Hu *et al.* where rod-like Zn-doped HA nanoparticles were observed [9]. Similar result was also found in the work of Qaisar *et al.* where rod-like Sr-doped HA particles were found [3]. This elongation associated with Sr doping is due to the substitution of Ca ions (180 pm) with Sr ions (200 pm).

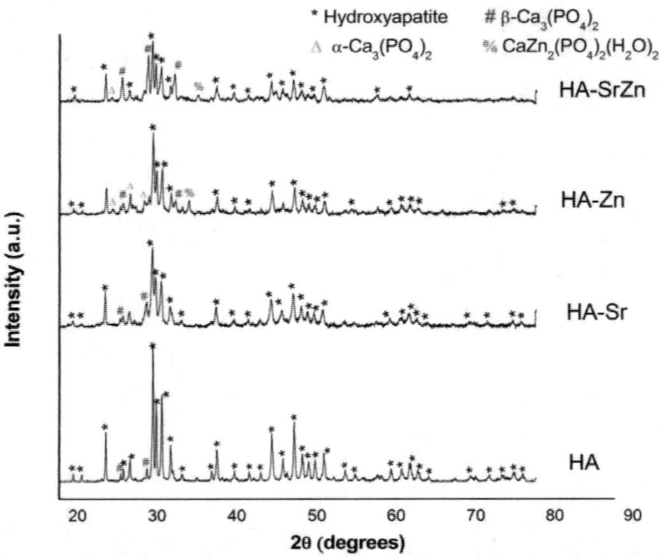


**Figure 1.** XRD pattern of HA, HA-Sr, HA-Zn and HA-SrZn powders.

### *3.2 Morphology and inner structure of electrospun scaffolds*

The SEM and TEM images of the electrospun fibrous scaffolds are shown in figure 3, respectively. From the SEM images, it can be seen that all the electrospun scaffolds exhibited a randomly interconnected and highly porous structure which resemble the extracellular matrix (ECM) of bone. The average diameter of the pure PCL fiber is approximately 612±3 nm. With an addition of HA particles, the average fiber diameters increase to 675±3, 674±2 669±3 and 687±1 nm, for the PCL/HA, PCL/HA-Sr, PCL/HA-Zn, and PCL/HA-SrZn composite fibers, respectively. All these sizes fall in the range of collagen fibers of bone which is 100–2000 nm [1] and enables the scaffolds with a structural similarity.

The surface of pure PCL fiber is relativity smooth, whereas those of the composite fibers are rough with particle protuberances. The TEM images reveals the inner structure of the composite fibers with the presence of the HA particles internalized or on the sub-surface of PCL fibers. This confirmed that the HA, HA-Sr, HA-Zn and HA-SrZn particles were successfully incorporated into the PCL matrix. However, agglomeration of the particles can be observed. This is likely because HA particles have low surface charge [8] resulting in a low electrostatic repulsive force between them.

## 4. Conclusions

In this study, Sr and Zn was doped or co-doped into the structure of hydroxyapatite by a sol-gel technique and characterized by the XRD and TEM. An electrospinning process was used to prepare the pristine PCL fiber and PCL/HA, PCL/HA-Sr, PCL/HA-Zn and PCL/HA-SrZn composite fibers. The composite fibers of PCL with doped HA may be expected to help enhancing the osteoblast growth in bone tissue regeneration.

## Acknowledgments

The authors would like to thank the Higher Education Research Promotion and National Research University Project of Thailand, Office of the Higher Education Commission, through the Advanced Functional Materials Cluster of Khon Kaen University.

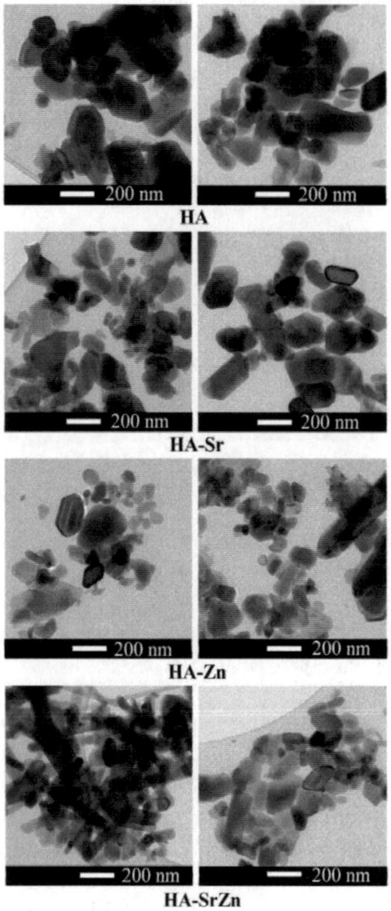


**Figure 2**. TEM images of HA, HA-Sr, HA-Zn and HA-SrZn powders.

**Figure 3.** (a) SEM images and (b) TEM images of PCL fiber and composite fibers.

**References**

[1] O'Brien F J 2011 *Mater. Today* **14** 88-95
[2] Orlovskii V P, Komlev V S and Barinov S M 2002 *Inorg. Mater.* **38** 973-984
[3] Qaisar S A, Bilton M, Wallace R, Brydson R, Brown A P, Ward M and Milne S J 2010 *J. Phys.: Conf. Series* **241** 012042
[4] Farzadi A, Bakhshi F, Solati-Hashjin M, Asadi-Eydivand M and Osman N A 2014 *Ceram. Int.* **40** 6021-6029
[5] Ravi N D, Balu R and Sampath Kumar T S 2012 *J. Am. Ceram. Soc.* **95** 2700-2708
[6] Kolmas J, Groszyk E and Kwiatkowska-Rozycka D 2014 *Biomed. Res. Int.* **15** 178123
[7] Cui A, Nelson B, Peng Y, Li K, Pilla S, Li W-J, Turng L-S and Shen C 2012 *Mat. Sci. Eng. C* **32** 1674–81
[8] Temprom L, Seet S L, Tippayawat P and Suwanna P. 2017 *Chiang Mai J. Sci.* **44** 630-639
[9] Hu W, Ma J, Wang J and Zhang S 2012 *Mat. Sci. Eng. C* **32** 2404-10

# Effects of chemical composite, puffing temperature and intermediate moisture content on physical properties of potato and apple slices

**S Tabtaing[1,*], S Paengkanya[2] and P Tanthong[1]**

[1]Department of Social and Applied Science, College of Industrial Technology, King Mongkut's University of Technology North Bangkok, Bangkok, Thailand
[2]Department of Sustainable Industrial Management Engineering, Faculty of Engineering, Rajamangala University of Technology Phra Nakhon, Bangkok, Thailand

*E-mail: Surapit.t@cit.kmutnb.ac.th; Surapit_t@yahoo.com

**Abstract.** Puffing technique is the process that can improve texture and volumetric of crisp fruit and vegetable. However, the effect of chemical composite in foods on puffing characteristics is still lack of study. Therefore, potato and apple slices were comparative study on their physical properties. Potato and apple were sliced into 2.5 mm thickness and 2.5 cm in diameter. Potato slices were treated by hot water for 2 min while apple slices were not treatment. After that, they were dried in 3 steps. First step, they were dried by hot air at temperature of 90°C until their moisture content reached to 30, 40, and 50 % dry basis. Then they were puffed by hot air at temperature of 130, 150, and 170°C for 2 min. Finally, they were dried again by hot air at temperature of 90°C until their final moisture content reached to 4% dry basis. The experimental results showed that chemical composite of food affected on physical properties of puffed product. Puffed potato had higher volume ratio than those puffed apple because potato slices contains starch. The higher starch content provided more hard texture of potato than those apples. Puffing temperature and moisture content strongly affected on the color, volume ratio, and textural properties of puffed potato slices. In addition, the high drying rate of puffed product observed at high puffing temperature and higher moisture content.

**Keywords:** puffing; chemical composite; texture; drying rate.

## 1. Introduction

A conventional drying method, hot air drying, is the simple method which produces the low crispness fruit chips. Moreover, this method is used a low temperature and longer drying time which provide the high shrinkage and hardness dried product. The high temperature and short drying time, puffing technique, is an alternative method for producing healthy crispy snack and can be improve the quality of product [1]. The puffing technique was applied successfully in various fruits and vegetables such as banana, potato, apple, carrot and elephant yam [2-4]. A high temperature is applied for accelerating the moisture content inside the food sample and suddenly evaporated in very short time [5, 6] This phenomenon leads to a rapid evaporation from inside the material to drying medium. As a result, the food structure was rapidly expanded, then porous structure was created therefore its texture become more crispness and low hard texture [7]. However, different crop variety showed different physical properties and chemical composites which significantly affect the puffed product quality. Tabtiang et al. [8] studied the effects of banana varieties on the quality of puffed banana. The results showed that the varieties of banana significantly affected on the product quality. The puffed Homtong-banana gave a lower hardness puffed banana than those from Namwa-banana, but poor in its colour quality.
Therefore, the objective of this work was to study the effects of puffing temperature and intermediate moisture content on the drying kinetic and qualities of puffed potato and apple slices. The product quality was evaluated in terms of colour, volume ratio and texture properties. In addition, the drying characteristics is also determined.

## 2. Materials and methods

### *2.1. Raw materials*

The raw materials for this experiment are potatoes and apples which were purchased from a local supermarket. They were washed and sliced to 2.5 mm thickness with 25 mm diameter. Potato slices were blanched by hot water at 95 ºC for 2 min while apple slices were not treated. Moisture content of initial and dried samples was followed from the AOAC method (1995).

### *2.2. Puffing process*

The puffing process consists of three drying steps. The first step, the sample slices were dried at a temperature of 90 ºC, air velocity of 3 m/s and 80% recycled air until the moisture content reached to 30, 40 and 50 % d.b. Then, the sample slices were puffed at a temperature of 130, 150 and 170 °C for 2 min The last step, those samples were dried again at 90 °C to the final moisture content 4 % d.b.

### *2.3. Colour measurement*

The colour of twenty puffed potato and apple slices were measured by a spectrophotometer (model ColorFlex, HunterLab Reston, VA, USA) with a D65 illuminate and observer angle of 10°. The colour of puffed samples were expressed in terms of L (lightness/darkness), a (redness/greenness) and b (yellowness/blueness).

### *2.4. Volume ratio measurement*

The volumes of both initial and puffed sample were determined by a solid replacement method by using glass breads [4]. The volume ratio of puffed product was described by Tabtiang et al. (2012).

### *2.5. Texture measurement*

The texture of puffed sample was evaluated by texture analyzer (TA-XT2i, Stable Micro Systems Texture, UK). The sample texture was evaluated in terms of hardness.

### *2.6. Statistical analysis*

The data from an evaluation of puffed product quality were analysed by the Analysis of Variance by using SPSS software (Version 17). Differences mean values were established using Tukey's Multi Range Tests at a confidence level of 95% ($p < 0.05$).

## 3. Results and discussion

### *3.1. Colour*

Figure 1 shows the colour of puffed potato and apple slices under different experimental conditions. It can be seen that the puffing temperature significantly affected colors of puffed samples. The color change ($\Delta E$) of both puffed products increase with increasing in the puffing temperature. This is because the high puffing temperature accelerated a non-enzymatic browning reaction [3, 10]. In addition, all cases of puffed potato slices gave more color change than that apple slices. However, the potato had lower amount of monosaccharide than apple slices as showed in Table 1. It may be because of enzymatic browning reaction in potato which more reactive faster than apple slices.

### *3.2. Volume ratio*

Figure 2 shows the effects of puffing temperature and intermediate moisture content on volume ratio of puffed potato and apple slices. It was found that the puffing temperature and moisture content significantly affected on the volume ratio of puffed potato slices. An increasing in puffing temperature and moisture content resulted in increasing volume ratio of puffed potato slices [3, 6]. The puffed potato slices show the higher volume ratio than those apple slices in all conditions. This is because potato sample had higher starch content than apple sample as shown in Table 1, resulting in large volume expansion of product structure [1]. In case of apple sample, both factors are not significantly affected on the textural properties of puffed apple slices.

### *3.3. Texture*

Figure 3 shows the effects of puffing temperature and intermediate moisture content on the hardness of puffed potato and apple slices. Increasing of puffing temperature and moisture content promoted lower hard texture of puffed potato. It is because the larger volume of potato sample as higher level of both factors provided more porous inside product structure therefore structural of puffed product become weaker [5, 7]. However, both factors were not significant effect on hardness of puffed apple slices. Considering between fruit variety, puffed potato slices had higher hardness than those apple slices. This is because of their amount of chemical composition such as starch of potato, had higher than apple sample.

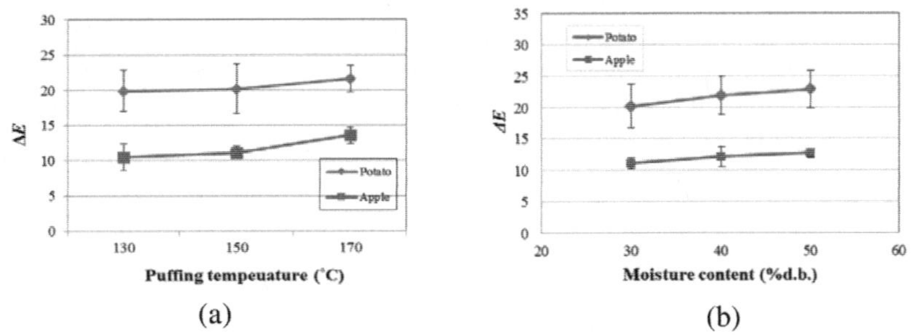


**Figure 1.** Effects of puffing temperature at moisture content of 40% d.b. (a) and intermediate moisture content at temperature of 150 °C (b) on $\Delta E$ of the dried potato and apple slices.

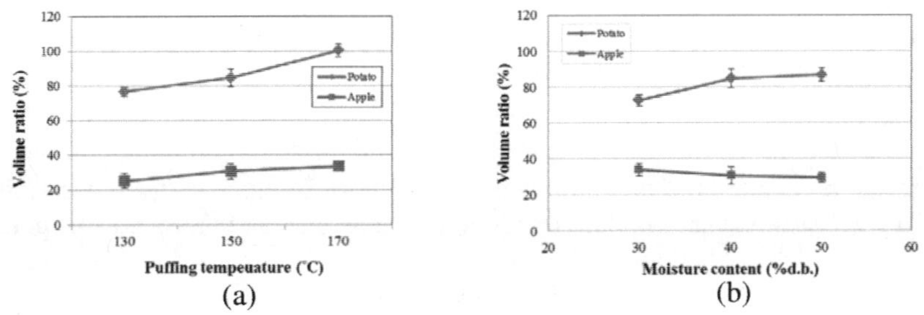


**Figure 2**. Effects of puffing temperature at moisture content of 40% d.b. *(a)* and intermediate moisture content at temperature of 150 °C (b) on volume ratio of the dried potato and apple slices.

**Table 1** Starch, some sugar and pectin content of fresh potato and apple.

| Fruits | Starch (% dry mass) | Fructose (% dry mass) | Glucose (% dry mass) | Pectin (% dry mass) |
|---|---|---|---|---|
| Potato | 17.50 | 0.20 | 0.20 | 0.7-1.5 |
| Apple | 0.05 | 6.00 | 2.00 | 0.78 |

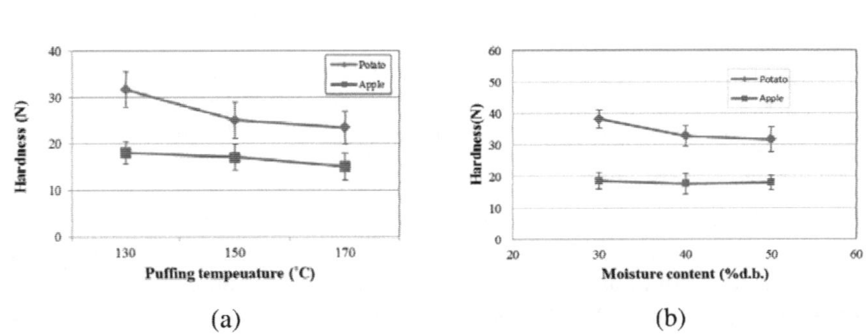


**Figure 3**. Effects of puffing temperature at moisture content of 40% d.b. (a) and intermediate moisture content at temperature of 150 °C (b) on hardness of the dried potato and apple slices.

### *3.4. Drying characteristics.*

The drying rates of potato and apple slice at different puffing temperature and intermediate moisture content is shown in Figure 4. The high puffing temperature and high moisture content of puffed potato products provided higher drying rate, resulting in a shorter drying time. This is because the puffed potato had higher volume then its moisture easy movement to its surface [1]. It should be note that, the drying rate of puffed apple slices was lower than puffed potato slices in all cases. It is due to the lower volume of apple than potato. In addition, higher amount of sugar content in apple can be react with moisture and then the moisture difficult to remove from inside apple tissue to its surface [8].

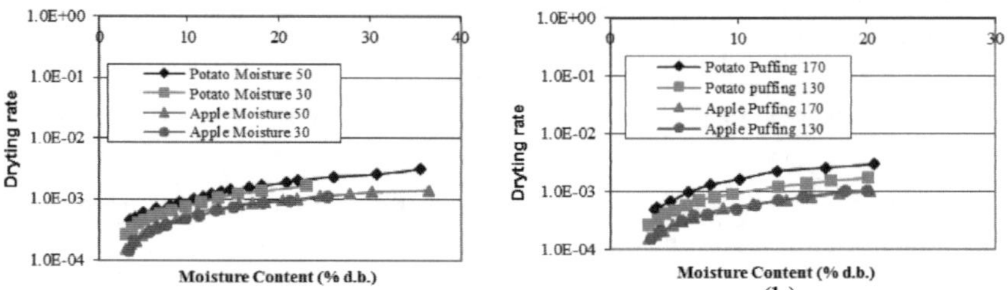


(a) (b)

**Figure 4.** Effects of (a) puffing temperature at 40% d.b. and (b) intermediate moisture content at 150 °C on the initial slope of the potato and apple slices.

## 4. Conclusion

Fruit variety affected on qualities of dried product. Potato could be puffed after processing therefore it had higher volume expansion than apple. However, the higher of starch content in potato provided harder texture and initial slope than apple. Moreover, higher puffing temperature and moisture content provided more color change, larger volume ratio, lower hardness of puffed potato slices. However, the effects of puffing temperature and intermediate moisture content had not significant effect on quality of puffed apple slices.

## 5. Acknowledgements

The research was found by King Mongkut's University of Technology North Bangkok "Contract no. KMUTNB-GEN-58-49" and College of Industrial Technology "Grant No. Res CIT 0302/2016", and The Drying Research Laboratory, School of Energy, Environmental and Materials, King Mongkut's University of Technology Thonburi for supporting of drying equipment in this project.

## 6. References

[1] Tabtiang S *et al* 2017 Optimum condition of producing crisp osmotic banana using superheated steam puffing *J. Sci. Food. Agri.* **97** 1244-1251

[2] Jayaraman K S *et al* 1980 Development of quick-cooking dehydrated pulses by high temperature short time pneumatic drying *J. Food Tech.* **15** 217-226

[3] Raikham C *et al* 2013 Optimum conditions of fluidized bed puffed for producing crispy banana *Drying Tech.* **31** 726-739

[4] Varnalis A I *et al* 2001 A proposed mechanism of high-temperature puffing of potato. Part I. the influence of blanching and drying conditions on the volume of puffed cubes *J. Food Eng.* **48** 361-367

[5] Raikham C *et al* 2015 Influence of pretreatments and drying process including fluidized bed puffing on quality attributes and microstructural changes of banana slices *Drying Tech.* **33** 915-925

[6] Hofsetz K *et al* 2007 Changes in the physical properties of bananas on applying HTST pulse during air-drying *J. Food Tech.* **83** 531-540

[7] Tabtiang S *et al* 2015 Optimal degree of ripeness and puffing temperature for the production of crisp Namwa and Homtong banana, *Thai Soc. Agri. Eng. J.* **23** 30-38
[8] Tabtiang S *et al* 2012 Effects of osmotic treatment and superheated steam puffing temperature on drying characteristics and texture properties of banana slices *Drying Tech.* **30** 20-28

# *Production of dried shrimp mixed with turmeric and salt by Spouted Bed technique enter the rectangular chamber.*

**P Thanthong*, Y Mustafa and D Ngamrungroj**

College of Industrial Technology, Department of Social and Applied Science, King Mongkut's University of Technology North Bangkok, Bangkok, Thailand
E-mail: pisut_2511@hotmail.com

**Abstract.** Today, dried shrimp in the market were refused food colour and drying until shrimp are colourful and tasty. Meanwhile, Community groups, women's health trying to produce food products come from herbs. As an alternative to consumers. The production process is also a traditional way to dry. In order to extend the shelf life longer. Sometimes, potential risks, both in quality and quantity of products. As a result, consumers are enormous. Thus, this research aims to study the possibility to produce shrimp dried mixed with turmeric and salt. Then dried shrimp mixed with turmeric and salt to keep up the quality criteria of the Food and Drug Administration-FDA It can reduce the risk of the consumer and can keep up in a kitchen Thailand. When buying shrimp from the fisherman's boat Will be made clear, clean impurities and shaking the sand to dry. Prepare a mixture of turmeric and salt. The shrimp were dipped into a beef with stirrer for 3 minutes. And scoop up centrifugal shrimp with dried. Measurement of initial moisture content averaging 78%wb. Then drying technique Spouted enter the rectangular chamber a continuous manner. Until average moisture content to 17%wb. The air temperature in the drying chamber at 180 $^{o}$C and hot air speed 4.5 m/s, a state heat transfer Mass and moisture within the shrimp. In chamber when drying, the shrimp have moved freely behaviour can spit water out faster does not burn. Shaving legs of shrimp shell fragments lightweight is sorting out the top of drying chamber. Private shrimp were dried out to the front of the quad drying chamber. Power consumption 27.5 MJ/kg, divided into electrical energy 12.3 MJ/kg and thermal energy is 15.2 MJ/kg. The hot air comes from burning LPG gas burner with dual automatic. And can adjustable to room temperature drying characteristics modulation setting.
**Key word** :Shrimp, Dried Prawn, Spouted bed, Heating chamber

## 1. Introduction

For dried shrimp, it was originally a shrimp for the family to live a family life. At this research, we offer shrimp from the sea fishing industry. Fresh shrimp from sea are a must. There is a need for a machine tool with a tactical and neat approach. To reduce the risk of loss of quality, volume of shrimp and price. Fresh shrimp from the fishing boat are frozen in the cold and the duration cannot be determined, but the average is about 10 days. When the boat is compared, the Marine Department will assign the fishing boat to loading and unloading in the loading area. Subsequently, raw materials in the ship will be transported to the designated channel. Then comes the various processes until the correctness of each criterion is determined. Then the raw materials past the sorting will be taken to the auction control room. To bid to buy the required. In the shrimp, Will be separated into groups and there are criteria to estimate how many impurities. Price will vary with impurity. Then sort the fresh shrimp and pack it in plastic bags of 20 kg each to facilitate the transfer to the dry shrimp factory as ordered by customers who are regular customers. When fresh shrimp crates are transported by small

IOP Conf. Series: Journal of Physics: Conf. Series **901** (2017) 012036 doi:10.1088/1742-6596/901/1/012036

trucks, they can control the temperature at 150°C. They will be checked by the quality control agency according to freshness criteria. Then issue SPECIMENT as a production order to the production unit. To prepare the material to be processed[1]. Shrimp is washed to free from fine sand particles. Then finely water and then turmeric for 90 seconds to 20 kg fresh shrimp weight. Delivered into the centrifugal dryer (drying by centrifugal force) for 150 seconds per weight fresh turmeric with turmeric 60 kg. Delivered to the dryer with a rectangular spouted bed dryer. The capacity is 300 kg/h until shrimp dried exothermic heat build-up comes close to atmospheric temperature. It will be put into plastic bags and paper boxes to facilitate the storage and transport, transport and transportation. This article is for research purposes, to improve the quality and efficiency of the dry shrimp process in the manufacturing industry. To enhance employee loyalty from factory manager, staff, practitioners and support agencies to have a quality manufacturing experience and performance for planning year. In 2018, Turmeric dried turmeric to Japan and England.

In the preliminary research study. Information is the company's information, as well as a collection of practical experience and produced as a product. For access to research and the data of the production process has allowed the drilling machine to be installed in the location. Employees will record all information for convenience. And it's easy to coordinate between real and current entities, the spec Shrimp of the shrimp, the time it takes to produce and the partner's purchasing power.

## 2. Machine and Method

### *2.1. rectangular spouted bed dryer*[2]

Shrimp dryer was made for dry shrimp trade. It will use LPG gas, generate heat and then reduce heat by fan. Hot air goes along a rectangular pipe through a sieve. It is a room between a fixed hot air room and a rectangular drying room with a dimension of 1.50 x 0.3 x 0.9 $m^3$. At the input position, it will use a set. Screws feeder and need to have a thread of 5 pitch conveyor screws all the time. Hot air does not come up as a barrier to enter dried shrimp into the drying room. 1 HP 380 V electric motor that can feed shrimp continuously at 200 kg/h. In the drying room[2], the hot steamed shrimp is sent to the bottom of the drying room. And a set of screws feeding the shrimp will be in contrast to the outlet of the dried shrimp. The screws are above the sieve plate for 18.3 cm shrimp and the dried shrimp leaves are removed from the drying chamber with a rectangular opening of 30x15 $cm^2$.There is a wind control board for the shuffling of shrimps that can be divided according to the criteria set to be free according to the amount of each shuffle, leaving the drying room to fall into the basket. The hot are has moisture content out of the drying chamber in the same vertical direction as the outlet of the dried shrimp. Shrimp shredding And the pulp bag slides down to the bottom of shrimp of the cyclone, falling into the prepared basket. The hot air will be forced out of the top of the cyclone into the atmosphere in the production room[3]. In the drying room there will be a ventilator to control the relative humidity in the room. And external dust to meet the criteria of the Department of Fisheries and provincial health control.

### *2.2. Drying of Shrimp*

This rectangle-shaped machine is designed and built to dry the cooked shrimp (Broiled Shirting).It has been around for years and the organization wants to launch new products. The objective of this study is to study the production of dried shrimp in the form of commercial trash. Shrimp was fed to dried with average humidity. 75%wb. It is fed into a rectangular drier using LPG for heat source generator. By using gas burners from Germany, which can heat up to 150 kW. Hot air pressurizes the hot air flowing through the rectangular pipe through the bottom of the sieve, supporting the hot compressed air, which flows through the pressure to flow across the drying chamber. The spatula shape is rectangular at a speed of 4.5 m/s (at the grill supports shrimp).The dimension of the rectangular spouted chamber is 30 x 150 x 147$cm^3$. The temperature inside the drying room is rectangular in shape is constant at 150 °C. Feeding the shrimp is adjusting feed rate and using default input values. 200kg / h and lobster when fed into the drying room. Shrimp behaviours started shifting from the beginning of the medium plate sieve to shrimp. And move to the front in a neat and independent manner, depending on the moisture content of each shrimp. Shrimp moves to the end of the sieve. Shrimp dried by the criteria will be out from the drying room[3,4]. The part of this position will be collected to analyze the results. This sample is then sent to the Quality Control Agency. To confirm by the order of the specification compliant partner. The average value of shrimp was 41.4%wb. The shrimp that were studied in this study were non-graded shrimp. However, it is a miniature product that contains dried by-products and dried by-products from the shrimps. Square spheres are as follows;

1. Dry Shrimp is the main product that the market needs.
2. Shrimp and shellfish out of the shrimp. While there is movement within the drying chamber.
3. Pulp from the shrimp that came out of the moustache and tail from shrimp.

## 3. Result

This experiment the study of drying fresh shrimp Miniature not graded. Feeding capacity 200 kg/h. The pressure of the hot air inside the drying chamber of the spouted bed is constant through drying shrimps processed working. Because the creation of a full state. It must be composed of two things that must be very closely related are hot air crossing the drying chamber. And the shredded raw prawns are free in the drying room. Both of these must be taken as a characteristic of the drying behaviour of the shrimp. Consistently meet the needs and goals of producing quality dried shrimp products.

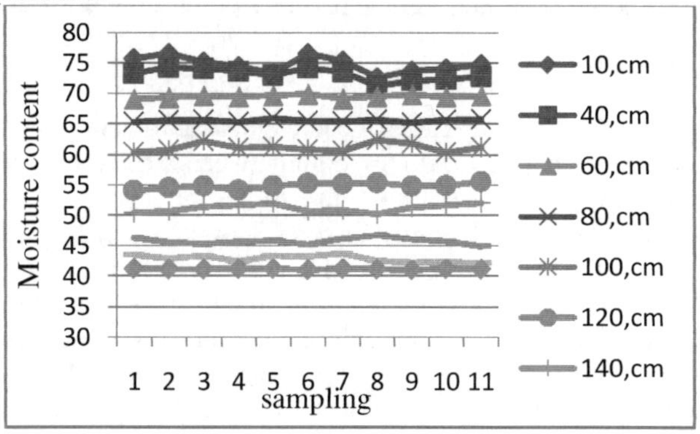


**Figure 1.** Graph moisture content of shrimp with sampling along drying chamber spouted bed

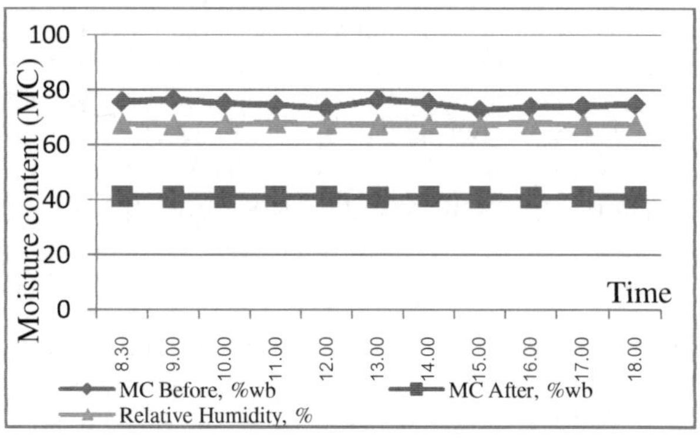


**Figure 2.** Drying of shrimp with continuous spouted bed technique this graph shows the difference in initial moisture content, which can be reduced to the final moisture content of the shrimp.

For the temperature of the hot air used in the drying room, effect of free shuffling behaviour of shrimp in the drying room show in Figure 1 and 2.When adjusting the hot air temperature at 150 $^{o}$C, The free shuffling of the steaming shrimp. The moisture content of shrimp did not affect the drying behaviour of shrimp. When the hot air temperature is raised to 180 $^{o}$C, the shifting of the shrimp does not change. Show that the lobster's popularity has no effect. But it requires more LPG gas. When the temperature inside the drying room is full, the bed is rectangular. And the structure of the dryer and auxiliary equipment. There will be more heat loss.

For the dynamics of drying It compares the change in moisture content with time of shrimp to move forward independently. In a full spouted bed inside a rectangular shaped chamber. The hot air that helps the shrimp move forward. Instantly following the timing of the spouted bed show in Figure 3.From Figure 3, the moisture content is rapidly reduced. Regularly after the moving shrimp move up to 30% of the full length of the drying chamber, the rectangular are filled with squares. The moisture content of the shrimp is consistent when the shrimp move to 90% of the length of the rectangular spouted bed. And at the end of the rectangular spouted bed. This provides a sufficiently dry area for the required final moisture content. Each shrimp will move rapidly according to the density of each shrimp. Then finally dry shrimp out of the rectangular spouted bed chamber.

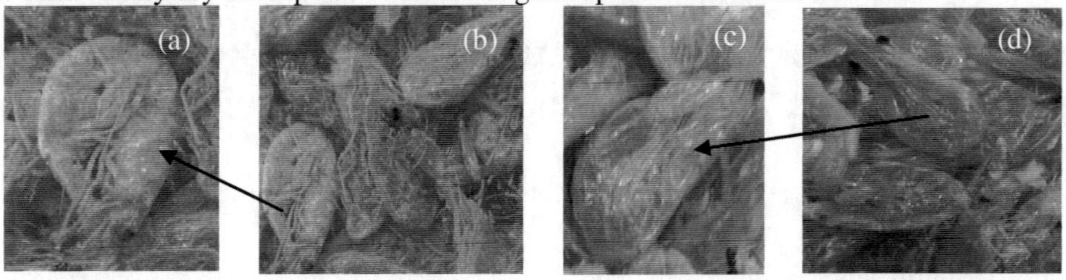


**Figure 3.** Continuous- spouted bed drying technique. From about 80%wb initial humidity. until the final moisture content is about 40%wb, with the eyes, legs, tentacles and shell of shrimp. Do not show burns due to heat. (a) and (b) Salted Turmeric, (c) and (d) Turmeric sweet turmeric

In comparison of the dynamics of drying and quality of dried shrimps produced from this rectangular spouted bed dryer. Before experimenting with a technique that emphasizes the characteristics of the shrimp. Shrimps were sampled to represent the drying by drying by baking. Electrically heated wire that adjusts to a temperature of 107 $^{o}$C and provide 76 hours of drying time. Based on the principle. According to Association of Official Agricultural Chemists 2000 (AOAC 2000) Food Analysis Standards. To be used as criteria in determining the distribution criteria of quality dried shrimp.

## 4. Conclusions

Summary of research results, Shrimp drying using rectangular spouted bed technique. This objective to produce shrimp sweet-salted turmeric with commercial quality. The spouted bed can set the initial moisture content of the shrimp. Shrimps through the process of fling with a mechanical mechanism until the value is less than 78%wb. Shrimp feeds into drying chamber, spouted bed, rectangular shape using hot air 150 $^{O}$C. The hot air flows across the grate plate so that the shrimp can move freely all the cross section of the great plate. Shrimp feeds with a capacity of 200 kg/h can be reduced to a final moisture content of 40.3%wb. The moisture ratio of the shrimps rapidly decreases in the rectangular spouted bed. Shrimp raw materials do not need to control size shrimps. The hot air velocity, temperature and the static pressure between the sieve plates must be constant. It can be produce turmeric products dried turmeric is a consistent colour. Although the Shrimp have different dimensions

## 5. Reference

[1]Yapha M., and soponsonnadit S., 1994 Design and testing of pilot continuous fluidized Bed Paddy dryer , waster at of Engineering, Energy technology king Mongkut's University of Technology Thonburi, pp.89-92

[2]Mustafa Y., Nathawood D.,2012, Rectangular plane dryer for paddy drying, the 11sd Conference on energy heat and mass transfer in thermal equipment and process, Haadchalao Chanthaburi, pp. 6-10

[3]Yapha M., Bunyawanichakul P., and Hayinilah N., 2014, coarse rice Products by must flow Fluidization techniques Diabetes Patents, International Journal at Science and Research ) IJSR(, issn: 2319-7064, Vol.3 issue 6, pp.5-10

[4]Yapha M., Bunyawanichakul P., and Hayinilah N.,2014, Must flow Dryer for Rough Rice the second international conference on green computing, technology and innovation, the Asia Pacific University of Technology and Innovation, Kualalumpur, Malaysia, March 18-20, pp. 26-30.

Siam Physics Congress 2017 (SPC2017) IOP Publishing
IOP Conf. Series: Journal of Physics: Conf. Series **901** (2017) 012037 doi:10.1088/1742-6596/901/1/012037

# Non-chemistry coupled PM10 modeling in Chiang Mai City, Northern Thailand: A fast operational approach for aerosol forecasts

**Ronald Macatangay**

National Astronomical Research Institute of Thailand, Chiang Mai, Thailand

**Gerry Bagtasa**

Institute of Environmental Science and Meteorology, University of the Philippines, Diliman, Quezon City, Philippines

**Thiranan Sonkaew**

Faculty of Science, Lampang Rajabhat University, Lampang, Thailand

ronmcdo@gmail.com

**Abstract**. The Weather Research and Forecasting (WRF v. 3.7) model was applied to model PM10 data in Chiang Mai city for 10-days during a high haze event utilizing updated land use categories from the Moderate Resolution Imaging Spectroradiometer (MODIS). A higher resolution meteorological lateral boundary condition (from 1 degree to 0.25 degree) was also used from the NCEP GDAS/FNL Global Tropospheric Analyses and Forecast Grid system. A 3-category urban canopy model was also added and the Thompson aerosol-aware microphysics parameterization scheme was used to model the aerosol number concentrations that were later converted to PM10 concentrations. Aerosol number concentration monthly climatology was firstly used as initial and lateral boundary conditions to model PM10 concentrations. These were compared to surface data obtained from two stations of the Pollution Control Department (PCD) of Thailand. The results from the modeled PM10 concentrations could not capture the variability ($r = 0.29$; 0.27 for each site) and underestimated a high PM10 spike during the period studied. The authors then added satellite data to the aerosol climatology that improved the comparison with observations ($r = 0.45$; 43). However, both model runs still were not able to capture the high PM10 concentration event. This requires further investigation.

## 1. Introduction

Seasonal blankets of haze cover Northern Thailand every dry season, which occurs from around mid-February to May. One of the main causes of this haze is the burning of agricultural waste from highland corn farms from commercial contract farming, majority of which belong to agricultural conglomerate companies, which sources corn from Thailand, Myanmar, Laos and Vietnam. This crop change, from rice, is a response to the increasing demand for animal feed and rising corn prices. It also yields quick earnings as corn can be harvested in four months. Other causes include forest burning by hunters to displace wildlife, vehicular emissions, urban sprawl, rubbish burn, harvesting of exotic mushrooms and enhanced biogenic volatile organic carbon (VOC) emissions due to land use change and increased temperatures.

Aerosols and particulate matter (PM) are some of the components of haze. They affect both health and climate. PM10, or particulates having a size of < 10 μm, increases the risk of tracheal, bronchial and lung cancer. Thailand also sets lower air quality standards than global benchmarks for PM10 at 120 μg/m3. The World Health Organization sets it at 50 μg/m3.

Aerosols also affect the climate by modifying cloud microphysical properties producing increased cloud albedo (first indirect effect) and reduced cloud droplet size that may delay and reduce precipitation (second indirect effect) [1]. Related to the second indirect effect is the phenomenon called aerosol enhanced conditional instability. In a polluted environment, with strong absorbing aerosols, greater amounts of moist static energy build up during daytime because convection is largely suppressed by the aerosol-radiation interaction compared to the clean case [2].

In February 2011, Oanh et al. analyzed the effect of meteorology and emissions for Chiang Mai City, during the March 2007 haze episode [3]. In the study, it was pointed out that due to the lack of adequate emission input data for this region required for chemical transport modeling to predict the haze, the climatological approach in combination with statistics were utilized. However, in December 2011, Amnuaylojaroen et al. was able to use the California Mesoscale Puff (CALPUFF) model, driven by the Weather Research and Forecasting (WRF) model, to investigate PM2.5 distributions as well as the relative contribution of fine fraction (PM2.5) and coarse fraction (PM10-2.5) to PM10 fraction from forest fires in the Chiang Mai basin also for March 2007 [4]. In the study, it was found out that strong atmospheric stability and light low-level winds over Chiang Mai were the favorable conditions for particulate matter accumulation.

In this work, the Weather Research and Forecasting (WRF) model [5] version 3.7 was utilized with the Thompson aerosol-aware microphysics scheme [1] to simulate PM10 concentrations in Chiang Mai City without chemistry coupling to provide a more computationally cost-effective model setup for potential aerosol operational forecasts.

## 2. Methodology

### *2.1. Numerical Weather Prediction*

In numerical weather prediction (NWP), physical laws of motion and conservation of energy that govern the evolution of the atmosphere (the dynamics) can be written into mathematical equations that can be solved numerically. For example,

$$\frac{\Delta A}{\Delta t} = f(A) \tag{1}$$

where $\Delta A/\Delta t$ is the change in a forecast variable, A, at a particular point in space with respect to time, t, and $f$(A) describes the physical processess that can cause changes in the forecast variable. Values of meteorological variables later in time are calculated by finding their initial values and then adding the physical forcing, $f$(A), that acts on the variables during the forecast period. This can be mathematically represented as,

$$A^{forecast} = A^{initial} + f(A)\Delta t \tag{2}$$

with the actual equations used are called the primitive equations. These equations dictate the forces or dynamics that give movement to air. It includes thermodynamic changes occurring in the atmosphere calculated from momentum, mass, energy and moisture conservation laws. Equally essential are: processes that occur at scales smaller than the model can resolve; energy, water and momentum exchanges between the atmosphere and other sources such as land, ocean and solar radiation; and cloud and precipitation physics. In NWP, the atmosphere is divided into smaller grid boxes (the size of which is called model resolution) where all these processes are calculated over an area which the model runs called the model domain [6].

*2.2. Dynamical Downscaling and Model Parameterization*
NWP models are usually driven by coarse spatial resolution (e.g. $1^{o}$ x $1^{o}$) global climate models that are unable to resolve sub-grid scale features (i.e. clouds, topography, etc.). In order to perform regional and local impact studies, downscaling has to be performed. One form is dynamical downscaling, where output from the coarse resolution model is used to drive a higher spatial resolution model. One such model which can perform dynamical downscaling is the Weather Research and Forecasting (WRF) model. This is used in this study.

However, even when dynamical downscaling is performed, there are still small-scale atmospheric processes that NWP models cannot resolve. The model must therefore account for the aggregate effects of these small-scale processes. This is called parameterization [7]. Some of the physical processes that are typically parameterized by models are condensation, turbulence, phase changes, surface roughness and deep convection. In order to produce fast aerosol simulations, parameterizing microphysics processes are essential.

Microphysics refers to the amount, type, processes (e.g. phase changes) and interaction between different condensates or hydrometeors (e.g. water vapor, cloud, rain, etc.). In this study, the Thompson aerosol-aware microphysics scheme was used. It was first utilized to investigate the effects of aerosols on cloud development and precipitation. It incorporates the activation of aerosols as cloud condensation nuclei (CCN) and ice nuclei (IN). In order to decrease the computational burden and complexity, the specific aerosol types as well as the chemical composition of multiple aerosol categories were simply referred to as "water-friendly" aerosols, or $N_{wfa}$ (hygroscopic aerosols) and as "ice-friendly" aerosols, or $N_{ifa}$ (non-hygroscopic ice-nucleating aerosols). $N_{wfa}$ is the sum of sulfates, sea salt and organic carbon which are related to PM10 concentrations.

Normally, PM10 modeling involves coupling to chemistry. However, coupling to chemistry proves to be computationally expensive as shown by the study of Wang et al. in 2010 wherein approximately 8000 computation hours per model month were needed [8]. This is due to the increased number of variables to simulate. It was also mentioned in [1], that the simplest WRF model with chemistry options available at that time increased the number of variables by over a factor of 2 massively impacting computer memory and time. This defeats the purpose of forecasting an event, which needs updating model runs for at least every 6 hours.

*2.3.Initial and Lateral Boundary Conditions, Model Domains and Validation Sites*
The model runs in this study used the NCEP GDAS/FNL 0.25 Degree Global Tropospheric Analyses and Forecast Grid lateral boundary conditions [9] with a 10 km outer domain (parent) covering most of Indochina and a 3.33 km inner domain (child) encompassing most of northern Thailand. WRF v. 3.7 was ran for the period of April 10-20, 2016 UTC coinciding with the time period before, during and after the Thai New Year. The Thai New Year, or Songkran, is a big event in Thailand with many tourists coming in to celebrate this holiday. It also coincides with the recurring biomass burning season in northern Thailand.

Since the validation sites for PM10 concentrations are in an urban environment and sensitive to surface winds, the MODIS land use dataset and a 3-categogy urban canopy model were utilized. The optimum cumulus parameterization obtained in [10], was also used. The rest of the parameters were kept at the WRF v. 3.6 defaults. The comparison sites for Chiang Mai city for PM10 concentrations are at the Pollution Control Department (PCD) air quality monitor stations located at the Chiang Mai City Hall and at the Yupparaj Wittayalai School.

Initial and lateral boundary conditions for the aerosol number concentrations were derived from multi-year (2001-2007) global simulations from the Goddard Chemistry Aerosol Radiation and Transport (GOCART) model including mass mixing ratios of sulfates, sea salts, organic carbon, dust and black carbon with 0.5 deg longitude by 1.25 deg latitude grid resolution [1]. This is designated as the control run. In order to incorporate up-to-date aerosol emission data, aerosol number concentrations derived from the Moderate Resolution Imaging Spectroradiometer (MODIS), on board Terra and Aqua satellites, were added to the aerosol climatology for the initial conditions. Aerosol number concentrations were then converted to PM10 concentrations using the method described by Mönkkönen et al. in 2004 [11]. This is designated as the experimental run.

## 3. Results and Discussion

After the simulations, the following model verification metrics [12] were used to assess the performance of the model: mean error, (multiplicative) bias, mean absolute error, root mean square error, mean squared error and correlation coefficient. A paired sample t-test (with 99% confidence interval) was also done to assess whether the performance between the control and the experimental model runs was significantly different.

In Figure 1, simulations using the 2001-2007 GOCART climatological and MODIS aerosol number concentration boundary conditions (referred to as "climatology" and "MODIS" hereafter, respectively) are compared to observations at the Chiang Mai city hall and at the Yupparaj Wittayalai PCD stations. The calculated model verification metrics are also shown in Table 1. For both sites, the MODIS model runs outperformed the climatology runs for all the verification metrics. A statistically significant difference was also obtained between the two datasets.

**Table 1.** Model verification metrics calculated after comparing the climatology and the MODIS model runs to PM10 observations at the Chiang Mai city hall and Yupparaj Wittayalai School PCD stations.

| | Perfect Score | Chiang Mai City Hall | | Yupparaj Wittayalai School | |
|---|---|---|---|---|---|
| | | Climatology | MODIS | Climatology | MODIS |
| Mean Error [microgram/m^3] | 0 | -62.20 | -52.57 | -79.20 | -69.24 |
| (Multiplicative) Bias, Bias | 1 | 0.49 | 0.56 | 0.41 | 0.48 |
| Mean Absolute Error, MAE [microgram/m^3] | 0 | 62.61 | 56.23 | 79.27 | 70.49 |
| Root Mean Square Error, RMSE [microgram/m^3] | 0 | 74.80 | 66.95 | 87.92 | 79.67 |
| Mean Squared Error, MSE [(microgram/m^3)^2] | 0 | 5595.70 | 4482.10 | 7729.90 | 6347.70 |
| Correlation Coefficient, r | 1 | 0.29 | 0.45 | 0.27 | 0.43 |
| **No. of Metrics Won** | | **0** | **6** | **0** | **6** |

Model outputs using MODIS captured more the variability at the two sites (r = 0.45; 0.43) as compared to the climatology run (r = 0.29; 0.27). However, both model runs failed to capture the high PM10 episode that occurred between April 16 – 17, 2016. This would have to be investigated in future work.

## 4. Conclusion and Recommendations

Weather models nowadays are usually characterized as having high resolution, fast computational speeds, relatively accurate and fairly simple representations of the state of the atmosphere in terms of meteorological variables such as temperature, pressure, relative humidity, rainfall, wind speed and wind direction. However, when considering air pollution concentrations, one normally couples these weather models with chemistry modules. Chemistry-coupled weather models are also characterized as having high resolution, but they are normally slow, have moderate accuracy and complicated. This study came up with PM10 simulations with speeds comparable to weather forecasts. This was attained by avoiding chemistry coupling, but instead utilized an aerosol-aware microphysics scheme combined with up-to-date satellite information of aerosol number concentrations for its initializations. The model results fairly captured the variability in the observations, but had difficulty simulating certain high PM10 events. This has to be studied further.

## References

[1] Thompson, G., and Eidhammer, T. (2014). A Study of Aerosol Impacts on Clouds and Precipitation Development in a Large Winter Cyclone. Journal of Atmospheric Sciences, 71, 3636–3658.

[2] Fan, J., Rosenfeld, D., Yang, Y., Zhao, C., Leung, L. R., & Li, Z. (2015). Substantial Contribution of Anthropogenic Air Pollution to Catastrophic Floods in Southwest. Geophysical Research Letters, 42(14), 6066–6075. http://doi.org/10.1002/2015GL064479

[3] Oanh, N. T. K., and Leelasakultum, K. (2011). Analysis of meteorology and emission in haze episode prevalence over mountain-bounded region for early warning. Science of the Total Environment, 409, 2261–2271. http://doi.org/10.1016/j.scitotenv.2011.02.022

[4] Amnuaylojaroen, T., and Kreasuwun, J. (2011). Investigation of Fine and Coarse Particulate Matter from Burning Areas in Chiang Mai , Thailand using the WRF / CALPUFF Investigation of Fine and Coarse Particulate Matter. Chiang Mai Journal of Science, 39(2), 311–326.

[5] Skamarock, W. C., J. B. Klemp, J. Dudhia, D. O. Gill, D. M. Barker, M. G Duda, X.-Y. Huang, W. Wang, and J. G. Powers, 2008: A Description of the Advanced Research WRF Version 3. NCAR Tech. Note NCAR/TN-475+STR, 113 pp. doi:10.5065/D68S4MVH

[6] Coiffier, J. Fundamentals of Numerical Weather Prediction, Cambridge University: Cambridge University Press: 2011, chapter 1.

[7] Laing, A., and J.-L. Evans, Introduction to Tropical Meteorology. Available from the NCAR MetEd website: http://www.meted.ucar.edu/tropical/textbook_2nd_edition/index.htm

[8] Wang, X., Liang, X., Jiang, W., Tao, Z., Wang, J. X. L., Liu, H., … Peckham, S. E. (2010). WRF-Chem simulation of East Asian air quality: Sensitivity to temporal and vertical emissions distributions. Atmospheric Environment, 44(5), 660–669. http://doi.org/10.1016/j.atmosenv.2009.11.011

[9] National Centers for Environmental Prediction/National Weather Service/NOAA/U.S. Department of Commerce. 2015, updated daily. NCEP GDAS/FNL 0.25 Degree Global Tropospheric Analyses and Forecast Grids. Research Data Archive at the National Center for Atmospheric Research, Computational and Information Systems Laboratory. http://dx.doi.org/10.5065/D65Q4T4Z. Accessed† 22 Apr 2016.

[10] Macatangay, R., Bagtasa, G. and Sonkaew, T., "Finding the Optimum Resolution, and Microphysics and Cumulus Parameterization Scheme Combinations for Numerical Weather Prediction Models in Northern Thailand: A First Step towards Aerosol and Chemical Weather Forecasting for Northern Thailand". International Symposium on Grids and Clouds 2016 Academia Sinica, Taipei, Taiwan. March 13-18, 2016.

[11] Mönkkönen, P., Umac, R., Srinivasanc, D., Koponena, I. K., Lehtinena, K. E. J., Hämeria, K., Kulmala, M. (2004). Relationship and variations of aerosol number and PM 10 mass concentrations in a highly polluted urban environment — New Delhi , India. Atmospheric Environment, 38, 425–433.

[12] Stanski, H.R., L.J. Wilson, and W.R. Burrows, 1989: Survey of common verification methods in meteorology. World Weather Watch Tech. Rept. No.8, WMO/TD No.358, WMO, Geneva, 114 pp.

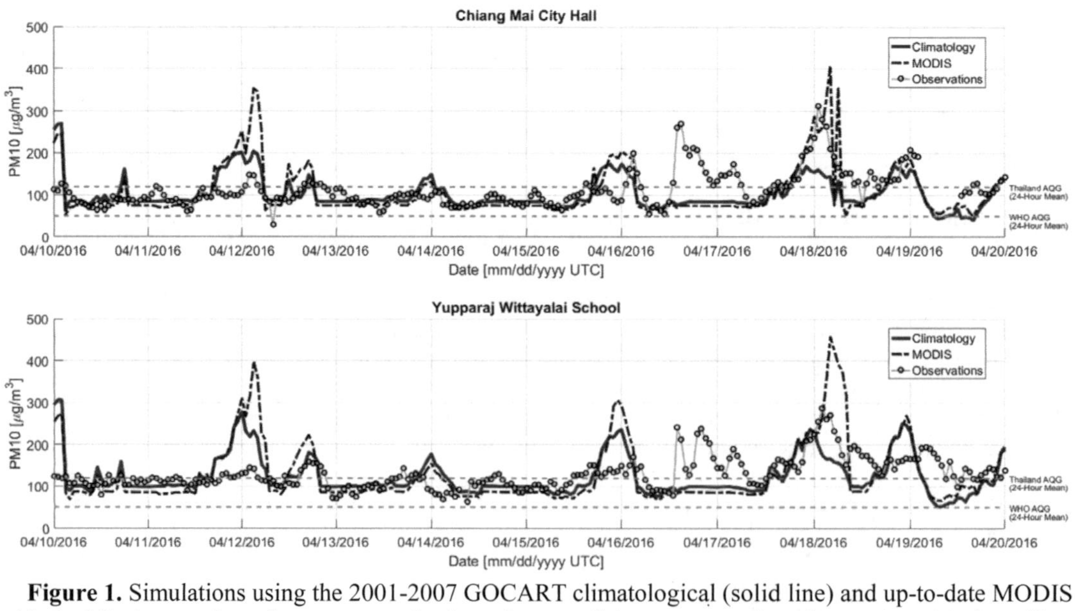


**Figure 1.** Simulations using the 2001-2007 GOCART climatological (solid line) and up-to-date MODIS (dashed line) aerosol number concentration boundary conditions compared to observations (circles with line) at the Chiang Mai city hall (upper panel) and Yupparaj Wittayalai School (lower panel) PCD stations (Note: the multiplicative bias calculated in Table 1 were applied for the model runs). Also shown are the World Health Organization (WHO) and the Thailand 24-Hour Mean Air Quality Guideline (AQG) value for PM10.

# Deep Shear Wave Velocity of Southern Bangkok and Vicinity

**T Wongpanit[1], K Hayashi[2], P Pananont[1,*]**

[1]SEIS-SCOPE, Department of Earth Science, Faculty of Science, Kasetsart University, Bangkok, Thailand
[2]Geometric Inc., San Jose, CA, USA

Email: fscipkp@ku.ac.th

**Abstract:** Bangkok is located on the soft marine clay in the Lower Chao Phraya Basin which can amplify seismic wave and can affect the shaking of buildings during an earthquake. Deep shear wave velocity of the sediment in the basin are useful for study the effect of the soft sediment on the seismic wave and can be used for earthquake engineering design and ground shaking estimation, especially for a deep basin. This study aims to measure deep shear wave velocity and create 2D shear wave velocity profile down to a bedrock in the southern Bangkok by the Microtremor measurements with 2 seismographs using Spatial Autocorrelation (2-SPAC) technique. The data was collected during a day time on linear array geometry with offsets varying between 5-2,000 m. Low frequency of natural tremor (0.2-0.6 Hz) was detected at many sites, however, very deep shear wave data at many sites are ambiguous due to man-made vibration noises in the city. The results show that shear wave velocity of the sediment in the southern Bangkok is between 100-2,000 $ms^{-1}$ and indicate that the bedrock depth is about 600-800 m, except at Bang Krachao where bedrock depth is unclear.

## 1. Introduction

Bangkok is located in the Lower Chao Phraya Basin which contains rather thick sediment which can amplify the seismic wave generated by an earthquake. Average shear wave velocity down to 30 meters ($V_{s30}$) in Bangkok have been measured in order to provide a shallow sediment property in earthquake engineering and earthquake ground motion estimation (i.e. Poovarodom, 2014). Although these studies have revealed relative low shear wave velocity on shallow layer of the entire, the shear wave velocities down to the bedrock which are important parameter to evaluate the basin effect are still absent. So far there are no direct and publically accessible evidence of the bedrock depth of Bangkok and vicinity. In addition, there are some contradiction reports about the depth to the bedrock in Bangkok area. Sinsakul (2000) proposed that the bedrock should be at depth between 500-2,000 m while Poovarodom and Jirasakjamroonsri (2014) suggested that the bedrock depths were between 400-800 m using microtermor measurement. This study aims to measure the deeper shear wave velocity in the southern Bangkok and make an attempt to identify the depth to bedrock of the study area.

SPAC (Spatial Autocorrelation) technique is considered to provide better resolution on dispersion image in the low frequency region and can generate a broader range of dispersion curve, compared to an F-K technique (Zhao and Li, 2010). In this study the two microtremor array (2-SPAC) technique was selected to measure the shear wave velocity down to 2 km at eight sites in the southern Bangkok and vicinity. The final results are displayed as shear wave velocity models.

## 2. Theories and techniques

Long period seismic wave from natural noise or microtremor are generated by natural phenomena such as ocean wave, atmospheric change and can be measured by microtremor observation. Spatial Autocorrelation was introduced by Aki (1957) using 4-7 sensors on triangle array

IOP Conf. Series: Journal of Physics: Conf. Series **901** (2017) 012038 doi:10.1088/1742-6596/901/1/012038

and one in its center. Assuming that microtremor propagates from all direction equally, Cho et al. (2006) and Margaryan (2009) showed that two-sensor SPAC yields almost identical phase velocities as those using 4 or 7 sensors. Hayashi and Underwood (2012) demonstrated that S-wave velocity profiles down to depth 2-3 km can be measured by using only 2 sensors. The phase velocity is calculated from the ambient noise using a formula:

$$\text{SPAC}\ (\Delta x,\omega) = J_0\left(\frac{\omega}{c(\omega)}\Delta x\right)$$

whereas x is a distance between receivers, ω is an angular frequency, c(ω) is a frequency-dependent phase velocity and $J_0$ is a zero order Bessel's function

The fundamental concept of SPAC is to record the vertical microtremor composition in several positions of the target sites to collect varied azimuthally angle propagation of Rayleigh wave identity. The data pair is calculated to compare its wave coherency spectrum. Then it conducts correlation, generates phase velocity and SPAC coefficient which fits to the formula above. The phase velocity model displays correlation of a velocity for each frequency. Then, it is converted to a 1D velocity-depth model by an inversion for the final result.

## 3. Data Acquisition and Processing

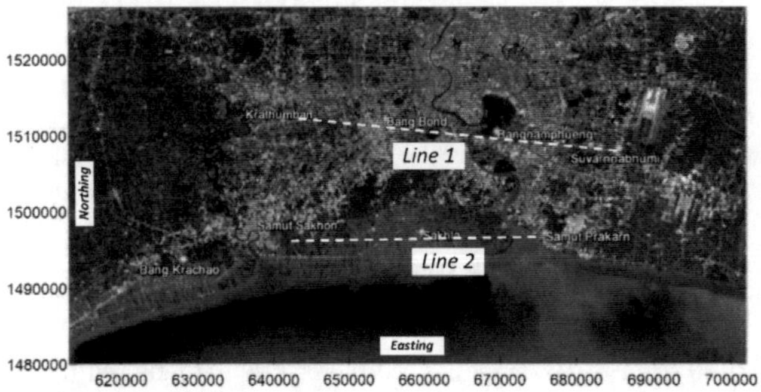


**Figure 1.** A map showing survey locations in the southern Bangkok. The cross-section of velocity profiles are shown in Figures 5 and 6.

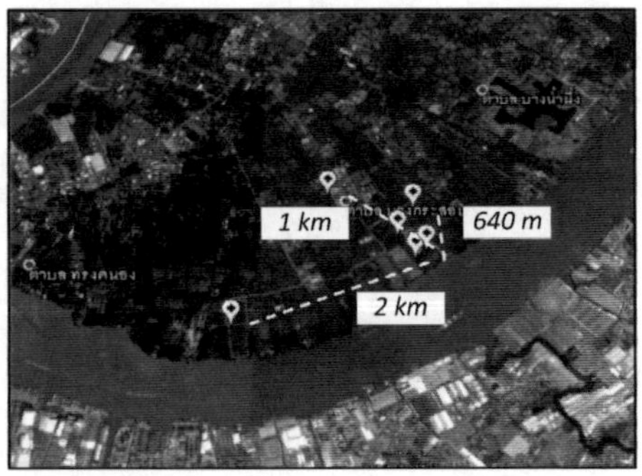


**Figure 2.** Example of a geometry of 2-SPAC methods at Bang Namphung site.

The survey locations were designed as line points to cover the southern Bangkok and its vicinity in quiet environment and less human activity. Inter-station location distance was about 15-18 km. Most observation areas are in parks, near abandoned buildings or agricultural lands. The data acquisition was done in day times during local winter period when the microseism from the Gulf of Thailand are strongest. Two MT-Neo seismographs (Geometrics Inc.) were placed as a linear array. One seismograph was fixed and acquired the data for entire time at each survey location. The second seismograph recoded the data at separation of 5, 10, 20, 40, 80, 160, 320, 640, 1,000 and 2,000 m with a record length of between 20-90 minutes or 2 hours at noisy site. The recording time were extended when the separation of two seismographs was increased. We used 2 ms sampling interval in this survey. The obtained data was divided into many blocks with about 160 seconds long for each block. Fast Fourier Transform (FFT) was used to convert these waveforms to frequency domain for calculating coherence of each data group. Summation of average complex coherence defined SPAC coefficient. Picked values of highest coherence are used for generating a phase-velocity relationship, aka. a dispersion curve. Then the shear wave velocity-depth models were generated by an inversion of dispersion curve. Figure 3 and 4 show phase-velocity images at Samut Sakhon and Samut Prakan sites. Their biggest coherence frequencies are between 0.5 and 0.2 Hz respectively. Red dots represent maximum coherence which are manually picked for generating a shear wave velocity-depth model by an inversion.

## 4. Results and Discussion

The phase velocity images show good coherence at minimum frequency of 0.2 - 0.6 Hz to maximum frequency of 5 Hz. Interference of the local long wavelength reduced clarity of deeper part of phase velocity image even though the possible longest wavelength obtained through the SPAC method is 2 to 4 times of receiver separation (Koichi, 2013). Joint Inversion (Suzuki and Yamanaka, 2010) were applied to all observed-selected dispersion curves to generate final 1D velocity-depth models. For the top 50 m depth, shear wave velocities are between 80-180 $ms^{-1}$. At deeper part, shear wave velocities range from 200-800 $ms^{-1}$ which could be sand and saturated clay layer that was found at depth between 100-500 m. On line 1, we found shear wave velocities of 1,000-2,000 $ms^{-1}$ at depth between 600-800 m: 600 m at Krathumban and 800 m at Bang Bond, Bang Namphueng and Savannabhumi.

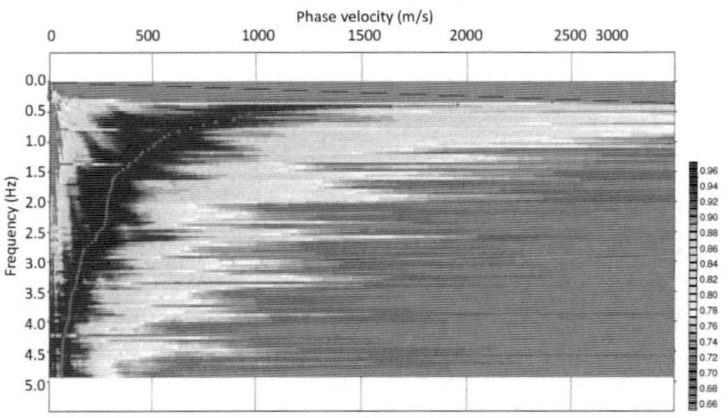


Figure 3. Phase velocity image at Samut Sakhon.

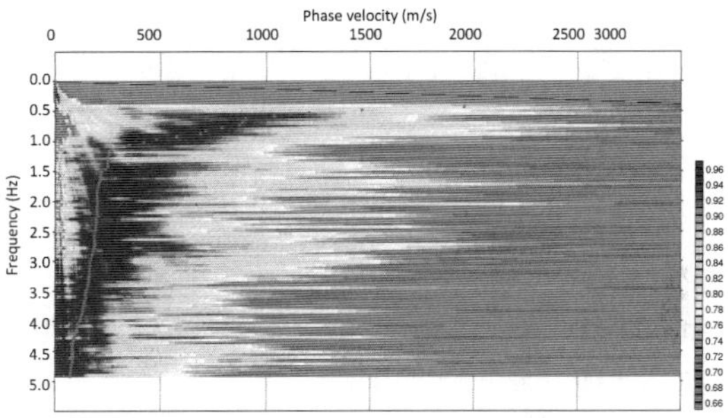


**Figure 4.** Phase velocity image at Bang Namphueng.

On line 2, the shear wave velocities of 1,000-2,000 $ms^{-1}$ were found at depth of 400 m at Bang Krachao, 700 m at Samut Sakhon, 800 m at Sakhla and 1,000 m at Samut Prakan. This velocity layer can be interpreted as sandstone and claystone bedrock which corresponds to Bangkok's geological structure (Sinsakul, 2000).

At the deeper level of some sites, we found the layer of shear wave velocity of 2,000-3,200 $ms^{-1}$ at depth of 1,500-1,900 m. It is expected to be quartzite basement layer. This shear wave velocity range was found in Krathumban at about 1,500 m depth. The same shear wave velocity range is also found at 1,700m depth in Bang Namphueng, Samut Sakhon and Sakhla and 1,900 m in Samut Prakan. This information conforms to the well data (Department Mineral Fuels, 1988).

There are some survey locations that the low frequency coherences are noisy due to location limitation of seismograph installation and surrounding activities. It affects the detection ability at deeper part, especially at basement level. The unclear deep velocity profiles are at Bangbon, Suvannabhumi and especially at Bang Krachao. The surface condition of these sites are soft soil, salt farm and near estate developing area.

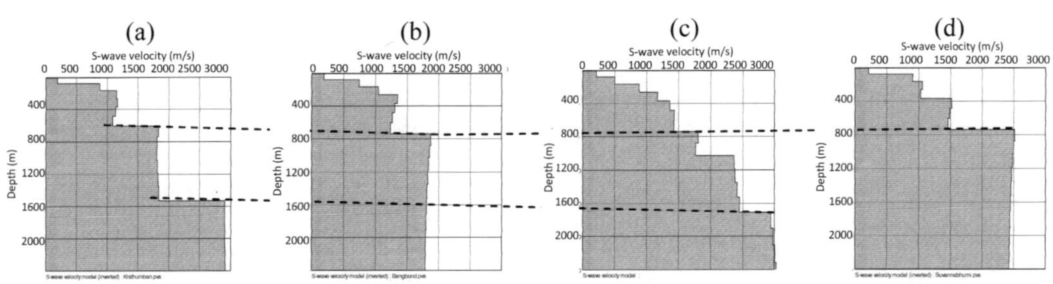


**Figure 5.** 1D velocity models of W-E transect line 1: (a) Krathumban (b) Bangbond (c) Bang Namphung and (d) Suvannabhumi respectively. The upper dashed line indicates the depth of the bedrock with shear wave velocity of 800-2,000 $ms^{-1}$ at depth of 700-800 m. The lower dashed line is a basement boundary with shear wave velocity of 2,000- 3,000 $ms^{-1}$ at depth of 1,500-1,700 m.

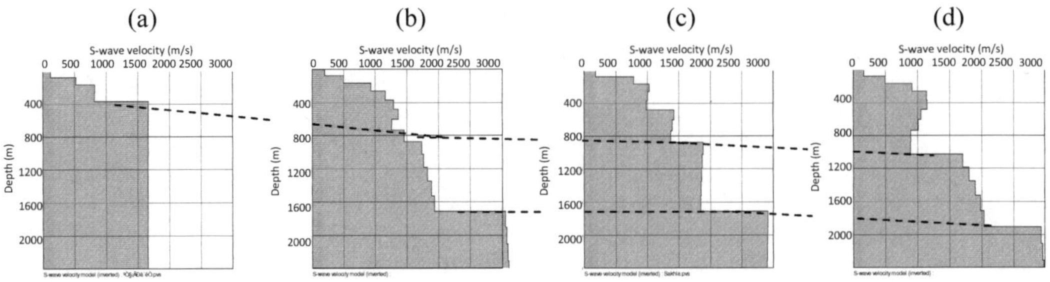


**Figure 6**. 1D shear wave velocity models of W-E transect line 2: (a) Bang Krachao (b) Samut Sakhon (c) Sakhla and (d) Samut Prakan respectively. The upper dashed line indicates bedrock depth level of shear wave velocity of 800-2,000 $ms^{-1}$ at depth of 400, 700-1,000 m. The second lower dashed line showing the boundary of basement with shear velocity of 2,000- 3,200 $ms^{-1}$ at depth 1,700-1,900 m.

## 5. Conclusion

Shear wave velocity profiles of the Southern Bangkok can be determined by a SPAC technique using 2-measurement station array (2-SPAC). Minimum value of 0.2 Hz phase velocity frequencies were found largely in most sites. The shear wave velocities of the topmost layer (50 m) are about 80-180 $ms^{-1}$. The shear wave velocities of the lower layers are interpreted as saturated sand, claystone and sandstone which is considered bedrock. Its depth varies from 600-800 m in the southern Bangkok, except at Bang Krachao. Basement depth with velocities of 2,000-3,200 $ms^{-1}$are found between depth of 1,500 to 1,900 m which corresponds to the well data.

One of the limitations of the data acquisition is strong noise interference of long wavelength which appeared in many areas. This interference reduces accuracy of the analysis or even tampers with all bedrock depth information. In the future, the 2-SPAC survey should be conducted throughout the Greater Bangkok with appropriate sites, spacing and correlate the result with data from other geophysical methods in order to create accurate bedrock map of the entire Greater Bangkok.

## Acknowledgement

We would like to thank the Faculty of Science, Kasetsart University for their financial support and Geometrics Inc. for supporting the seismographs. We also like to thank all assistance their help in collecting data.

## References

[1] Zhao D and Li V 2010 Comparison of FK and SPAC Methods in Determining Dispersion Curves from Passive Surface Waves. *Sym. on the App. of Geophys. to Eng. and Envi. Problems* 2010: 536-542.

[2] Sinsakul S 2000 Late Quaternary geology of the Lower Central Plain, Thailand. *J Asian Earth* Sciences **18** 415-426

[3] Tuladhar R et al. 2004 Seismic Microzonation of the Greater Bangkok Area Using Microtremor Observations. *Earthquake Eng. and Struct. Dyn.* **33** 211–225

[4] Yokoi T et al. 2009 Exploration Methods Using Microtremor Array. *Proc. of the 9th SEGJ Int. Symp.*– Imaging and Interpretation

[5] Aswad S et al. 2012 Bedrock Effect on Estimating Sediment Thickness Using Microtremor Analysis Methods. *Istanbul Int. Geophys. Conf. and Oil & Gas Exhibition*, Istanbul, Turkey

[6] Poovarodom N and Jirasakjamroonsri P 2015 Investigation of Long Period Amplification in the Greater Bangkok Basin by Microtremor Observations. *Proc. of the Tenth Pac. Conf. on Earthquake Eng. Build. an Earthquake-Resilient Pacific*

[7] Department Mineral Fuels 1988 Well Logging Data Well B

IOP Conf. Series: Journal of Physics: Conf. Series **901** (2017) 012039 doi:10.1088/1742-6596/901/1/012039

# Comparison of spectral ultraviolet irradiance measured from satellite and ground-based instrument at Nakhon Pathom province

**J Sriwongsa* and S Buntoung**
*Solar Energy Research Laboratory, Department of Physics, Faculty of Science, Silpakorn University, Nakhon Pathom, 73000, Thailand

*E-mail: muaytabo@gmail.com

**Abstract.** In this study, comparisons of spectral ultraviolet irradiance at 305, 310, 324 and 380 nm at the overpass time retrieved from OMI/AURA satellite with that from ground-based measurements were performed at Nakhon Pathom (13.82ºN,100.04ºE), Thailand. The analyzed data period comprised from 1 January 2010 to 31 December 2015. The comparison results clearly showed the overestimation of satellite data with root mean square difference (RMSD) between 22.9 and 48.9%, and mean bias difference (MBD) between 5.3 and 39.8% for all sky conditions, and reduced to 10.6-40.5% and 0.18-34.9% for clear sky conditions. Further results showed that the differences between the two datasets depend on atmospheric aerosol loads and clouds.

## 1. Introduction

The solar ultraviolet (UV) radiation is a small part of electromagnetic radiation which its wavelength ranges from 100-400 nm. Due to the biological effect, the UV radiation can be divided into three types; UV-A (315-400 nm), UV-B (280-315 nm) and UV-C (100-280 nm). Only UV-A and UV-B can reach the earth surface while UV-C is mostly absorbed by stratospheric ozone and cannot incident the surface. UV-A and UV-B have both positive and negative impacts on human life, animals and plants. For example, UV-B can synthesis vitamin D in human skin and by contrast, it can cause skin cancer. For these reasons, surface UV irradiance data is very important. This data can be retrieved from ground-based and satellite measurements. Although the ground-based measurement is the most accurate solution for obtaining UV radiation data, ground-based measurement is very scarce as it is costly. Therefore, satellite data is an alternative approach as it can provide the data globally. The validation of satellite data have been performed in several studies. For example, the validation of the daily erythemal UV doses from satellite products [1] and spectral UV radiation [2] were carried out in Europe. For Thailand, the differences of the UV index [3] and also erythemal UV doses [4] were presented. Thus, in this study, spectral UV irradiance at 305, 310, 324 and 380 nm from ground-based measurement and Ozone Monitoring Instrument (OMI) product in a tropical environmental of Thailand were compared. In addition, effects of cloud and aerosol on the difference of the two datasets were also analyzed.

## 2. Measurement and Data

### *2.1. Ground-based data*
The ground-based spectral UV irradiance data used in this study were measured by a spectroradiometer (Bentham, model DMc150) installed at Faculty of Science, Silpakorn University, Nakhon Pathom (13.82°N, 100.04°E), Thailand, as shown in figure 1. This instrument was used to measure the spectral irradiance from 260-420 nm with a resolution of 1 nm.

**Figure 1.** The Bentham spectroradiometer installed at Faculty of Science, Silpakorn University, Nakhon Pathom, Thailand.

The signal retrieved every 10 minute was recorded by using a PC. The signal was then converted into spectral UV irradiance using its sensitivity as follows:

$$I_\lambda = \frac{c_\lambda}{s_\lambda} \tag{1}$$

where $I_\lambda$ is spectral UV irradiance (mW/m$^2$/nm), $c_\lambda$ is signal at each wavelength (nA) and $s_\lambda$ is sensitivity of each wavelength (nA/nm/mW/m$^2$). The spectral UV data at 305, 310, 324 and 380 nm matching the OMI/AURA satellite product were averaged within ±30 minute of the OMI overpass time during 1 January 2010 to 31 December 2015.

Apart from the spectral UV data, aerosol optical depth (AOD) at 340 nm measured using a sunphotometer (Cimel, model CE318) installed at the site was collected from the Aerosol Robotic Network (AERONET) website. Cloud transmission factor (CLT) was also calculated as follows:

$$CLT = \frac{UV_{measured}}{UV_{clear}} \tag{2}$$

where $UV_{measured}$ is UV irradiance at 340 nm measured by the Bentham spectroradiometer under actual sky conditions and $UV_{clear}$ is UV irradiance at 340 nm under clear sky condition calculated from UVSPEC radiative transfer model. In addition, sky conditions were classified by using sky images taken by a sky view (Prede, model PSV100) installed at the site during the five year period.

### *2.2. Satellite data*
OMI is one of instruments onboard the NASA EOS/AURA satellite which is a sun-synchronous polar orbit. It measures reflected and backscattered solar radiation in the range of 270-500 nm, with a spatial resolution of 13 km×24 km at nadir. One of the OMI/AURA products is overpass time spectral irradiances at 305, 310, 324 and 380 nm, which is around 1 pm - 3 pm (local time) of the studied site. These data from 1 January 2010 to 31 December 2015 were retrieved and used to compare with the ground-based data.

## 3. Methodology
To validate the OMI spectral UV product, the OMI overpass time data at 305, 310, 324 and 380 nm were plotted against those from the ground-based data under clear sky and all sky conditions. Then

root mean square difference (RMSD) and mean bias difference (MBD) were calculated to determine the differences between the two datasets.

For the study of the influences of the atmospheric parameters, the ratios of the ground-based data to the satellite data for the four wavelengths were calculated. If the ratio equals 1, it means that the satellite data is equal to the ground-based data. Then the ratio were plotted with the aerosol optical depth for clear sky conditions and with the cloud transmission factor for all sky conditions. Finally, the changes of the ratios for each atmospheric parameter were analyzed.

## 4. Result and discussion

The comparisons between the spectral UV at the four wavelengths from OMI and ground-based measurements were shown in figure 2. It can be seen that the OMI data overestimated the ground-based data for all wavelengths with the RMSD ranging from 22.9-48.9% and MBD from 5.3-39.8% for all sky conditions, and 10.6-40.5% and 0.18-34.9% for clear sky conditions.

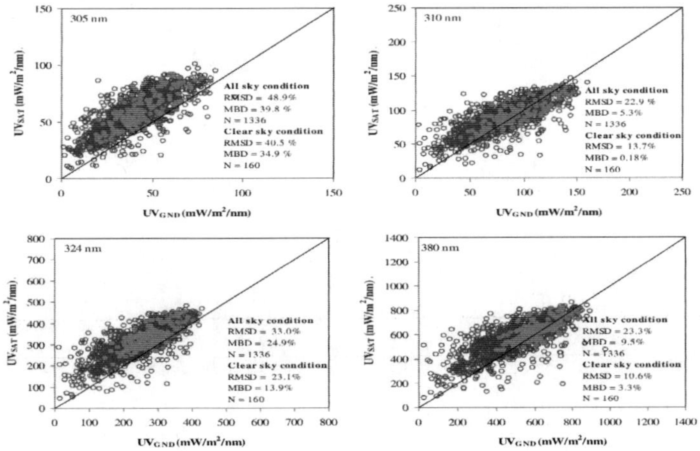


**Figure 2.** Comparison between the satellite-based ($UV_{SAT}$) and ground-based ($UV_{GND}$) overpass time spectral UV under clear sky (•) and all sky conditions (o).

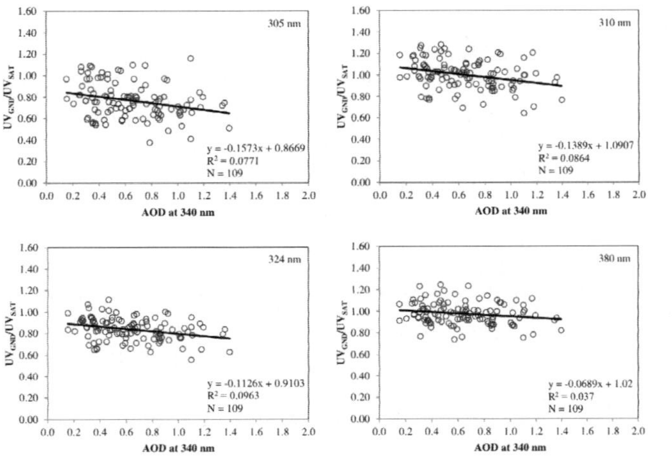


**Figure 3.** Relationship between the ratio of ground-based ($UV_{GND}$) to satellite ($UV_{SAT}$) spectral UV at overpass time and the aerosol optical depth.

These differences may be caused by the mismatches of time and space of the two datasets. In addition, the overestimation of the satellite data can be obtained as a result of cloud and aerosol contamination which can be seen in figure 3 and figure 4. The relationship between the ratio of the ground-based data to the OMI data and the aerosol optical depth for each wavelength under clear sky conditions was shown in figure 3. The ratio decreases from 1 to 0.5 when the atmospheric aerosol increases especially for shorter wavelengths. For the influence of cloud, figure 4 presented that when it is no cloud (CLT equals one), the OMI data is equal to the ground-based data, while the ground-based data is almost half of the OMI data when it is overcast

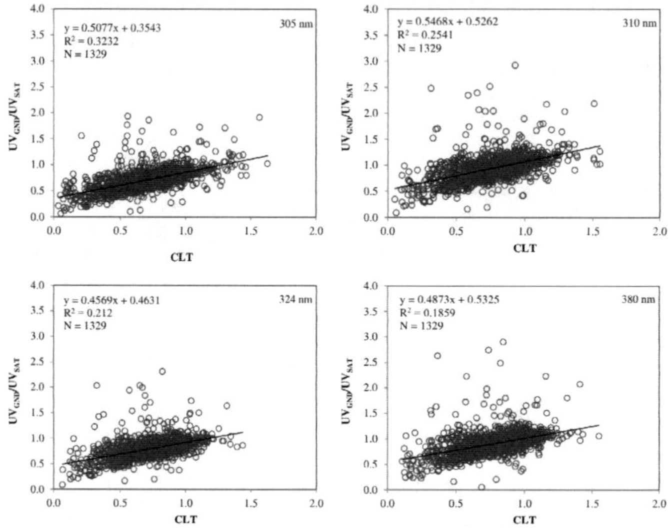


**Figure 4.** Relationship between the ratio of ground-based to ($UV_{GND}$) satellite ($UV_{SAT}$) spectral UV at overpass time and the cloud transmission factor.

## 5. Conclusion

The comparisons of overpass time spectral UV at 305, 310, 324 and 380 nm from OMI/AURA satellite and the Bentham spectroradiometer at Nakhon Pathom, Thailand were analyzed. The results showed the overestimation of the OMI data for all wavelengths, even under clear sky conditions. Thus, in order to correct the satellite-derived UV data, the influences of aerosols and clouds on the UV data should be examined. Finally, The linear relationships between the ratio of the satellite-derived to measured UV and aerosols and clouds were obtained as a result.

## References

[1] Tanskanen A, Lindfors A, Määttä A, Krotkov N, Herman J, Kaurola J, Koskela T, Lakkala K, Fioletov V, Bernhard G, McKenzie R, Kondo Y, O'Neill M, Slaper H, den Outer P, Bais A F and Tamminen J 2007 *J. Geophys. Res.* **112** D24S44

[2] Kazadzis S, Bais A, Arola A, Krotkov N, Kouremeti N and Meleti C 2009 *Atmos. Chem. Phys.* **9** 585–594

[3] Janjai S, Wisitsrikum S, Buntoung S, Pattarapanitchai S, Wattan R, Masiri I and Bhattarai B K 2014 *Int. J. Climatol.* **34** 453-461

[4] Buntoung S and Webb A R 2010 *J. Geophys. Res.* **115** D1821

IOP Conf. Series: Journal of Physics: Conf. Series **901** (2017) 012045 doi:10.1088/1742-6596/901/1/012045

# Determining the appropriate altitude to improve accuracy in rainfall estimation from radar reflectivity data

**M Lakkham**

Department of Earth Sciences, Faculty of Science, Kasetsart University, Bangkok, 10900, Thailand

E-mail: mathlakkham@gmail.com

**Abstract.** In present day, rainfall estimation by weather radar is widely used. Meteorologists use various z-r relationships to appropriate rainfall estimation for different study areas which have different factors such as topography, climate, rain pattern, types of clouds, etc. This research aims to determine the appropriate altitude of radar reflectivity (Z-Level) to provide accurate rainfall estimation. This research will use radar reflectivity data (Z) from Omkoi radar station, which is owned by the Department of Royal Rainmaking and Agricultural Aviation (DRRAA). It uses radar reflectivity data from 1.5 to 13.5 kilometers above mean sea level. We convert it to daily rainfall from radar (R) using Omkoi Z-R relationship (Z=92.4R^1.5) and compare the result to daily rainfall measured by rain gauge stations (G) from 82 rain gauge stations of Thai Meteorological Department (TMD), covering areas within radius of Omkoi radar station (180 km) in the northern of Thailand. The study will consider comparing various statistics of rainfall from radar (R) and rainfall from rain gauges (G), such as correlation coefficients, root mean square error (RMSE) and the characteristic distribution of graphs (Scatter Plot). The results show that the radar reflectivity data in altitude 3.5 km above mean sea level are the most suitable to be used to determine the Z-R Relationship. Therefore, determination of Z-R relationship of Omkoi radar station should be based on the radar reflectivity data at altitude 3.5 km above mean sea level.

## 1. Introduction

In present day, rainfall estimation by weather radar is widely used. Meteorologists use various z-r relationships to appropriate rainfall estimation for different study areas which have different factors such as topography, climate, rain pattern, types of clouds, etc. This research aims to determine the appropriate altitude of radar reflectivity (Z-Level) to provide accurate rainfall estimation.

Radar reflectivity in each altitude have relationship with rainfall measured by rain gauge. Relationship will vary according to the distribution of reflectivity, which affects the accuracy of radar rainfall estimation and each radar station has different appropriate altitude. Therefore, determining the appropriate altitude is important. It should have a proper altitude before. It will make an assessment of rainfall radar station that has the highest precision.

In this research, the criteria used in determining the proper altitude will consider various statistics, such as root mean square error (RMSE), pearson correlation coefficient, mean absolute error (MAE), bias and consider the distribution graph of rainfall from the radar (R) and rainfall from rain gauge (G) (scatter plot).

## 2. Literature review

### *2.1. Radar reflectivity factor*

The reflectivity factors at horizontal and vertical polarizations are given by:

$$Z = \frac{\lambda^4}{\pi^5 |K|^2} \int \sigma(D) N(D) dD$$

Where D = drop diameter
$\lambda$ = radar wavelength
$|K|^2$ = refractive index of the hydrometeors (approximately 0.93 for water and 0.20 for ice)
$\sigma(D)$ = backscattering cross section of scatterers
$N(D)$ = drop size distribution

If the scatterers are considered as small water spheres with small radii compared to the radar wavelength, then the approximation of the Rayleigh scattering applies and Equation can be expressed as:

$$Z = \int D^6 N(D) dD$$

### *2.2. Z-R Relationship*

Weather radar can effectively provide high resolution either spatial and temporal rainfall estimation. Rainfall estimation using weather radar in areas with insufficient rainfall stations. Weather radar cannot be used to measure rainfall intensities directly, so relationship between radar reflectivity data (*Z*) and rainfall rate (*R*), called the *Z-R* relationship ($Z = aR^b$) is used to estimate rainfall intensities using radar. (Mapiam and Sriwongsitanon, 2008)

The Z-R relationship is represented in term of empirical power law equation as below:

$$Z = aR^b$$

Where Z = radar reflectivity ($mm.^6/m.^3$)
R = rainfall intensity (mm./hr.)
a and b = parameters

### *2.3. Statistical measures*

In this study, we use 4 various statistical values to determine the appropriate altitude of radar reflectivity data for rainfall estimation.

*2.3.1. Mean Absolute Error (MAE).* In statistics, mean absolute error (MAE) is a measure of difference between two continuous variables. Assume X and Y are variables of paired observations that express the same phenomenon. Examples of Y versus X include comparisons of predicted versus observed, in this study we use R and G, R is predicted and G is observed.

*2.3.2. Root Mean Square Error (RMSE).* RMSE is a frequently used measure of the differences between values predicted by a model or an estimator and the values actually observed.

*2.3.3. Bias.* Statistical bias is a feature of a statistical technique or of its results whereby the expected value of the results differs from the true underlying quantitative parameter being estimated.

*2.3.4. Pearson correlation coefficient.* Pearson correlation coefficient is a measure of the linear correlation between two variables X and Y. It has a value between +1 and −1, where 1 is total positive linear correlation, 0 is no linear correlation, and −1 is total negative linear correlation. It is widely used in the sciences.

## 3. Methodology

### *3.1. Data*

This research will use radar reflectivity data (Z) from Omkoi radar station, which is owned by the Department of Royal Rainmaking and Agricultural Aviation (DRRAA). It uses radar reflectivity data from 1.5 to 13.5 kilometers above mean sea level and convert to daily rainfall data from radar (R) by using Omkoi Z-R relationship (Z=92.4R^1.5) and use daily rainfall data measured by rain gauge stations (G) from 82 rain gauge stations of Thai Meteorological Department (TMD), covering areas within radius of Omkoi radar station (180 km) in the northern of Thailand. (**Figure 1.**)

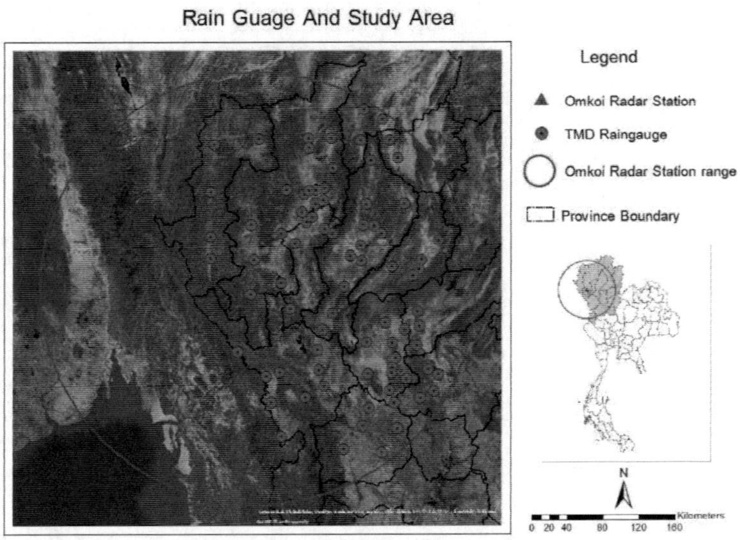


**Figure 1.** Rain gauge and study area map

*3.2. Select altitude*

After matching radar reflectivity and rainfall, will then calculate and compare various statistics of rainfall from radar (R) and rainfall from rain gauges (G), such as correlation coefficients, root mean square error (RMSE) and the characteristic distribution of graphs (Scatter Plot).

After calculate the statistics, select data by select by date and station which have correlation coefficient more than average and select altitude (Z-Level) by the statistics which have error less than average and consider the scatter plot of rainfall from the radar (R) and rainfall from rain gauge (G).

## 4. Result

*4.1. Statistical comparison between before and after the selected altitude.*

**Table 1.** Statistics of dataset before and after selected altitude.

| Data | N | Pearson | MAE | RMSE | Bias |
|---|---|---|---|---|---|
| Before (All Z-Level) | 23997 | 0.27 | 10.9 | 17.5 | 1.22 |
| After (Z-Level 2.5-5.5) | 5691 | 0.57 | 9.1 | 14.9 | 0.82 |

*4.2. Statistical comparison between Z-R relationships obtained from data before and after the selected altitude.*

**Table 2.** Statistics of Z-R relationship obtained from data before and after the selected altitude.

| ZR | Pearson | MAE | RMSE | Bias |
|---|---|---|---|---|
| Z-R Before (a=30.603, b=1.5691) | 0.27 | 15.96 | 27.73 | 0.68 |
| Z-R After (a=105.06, b=1.5282) | 0.57 | 8.15 | 12.98 | 0.95 |

*4.3. Statistical comparison between each selected altitude.*

*4.3.1. Use Omkoi Z-R relationship (Z=92.4R^1.5)*

**Table 3.** Statistics of data from each altitude use Omkoi Z-R relationship.

| Z-Level | Pearson | RMSE | Bias |
|---|---|---|---|
| 2.5 | 0.58 | 11.56 | 1.03 |
| 3 | 0.58 | 11.45 | 1.02 |
| 3.5 | 0.58 | 11.37 | 1.01 |
| 4 | 0.58 | 11.64 | 0.98 |
| 4.5 | 0.58 | 11.66 | 0.96 |
| 5 | 0.58 | 11.68 | 0.96 |
| 5.5 | 0.58 | 13.49 | 0.79 |

*4.3.2. Use Z-R relationship obtained from each altitude.*

**Table 4.** Statistics of data used Z-R relationship obtained from each altitude.

| Z-Level | Pearson | RMSE | Bias |
|---|---|---|---|
| 2.5 | 0.41 | 13.5 | 1.02 |
| 3 | 0.42 | 13.3 | 1.01 |
| 3.5 | 0.43 | 13.2 | 1.01 |
| 4 | 0.42 | 13.6 | 0.99 |
| 4.5 | 0.42 | 13.6 | 0.99 |
| 5 | 0.42 | 13.6 | 0.98 |
| 5.5 | 0.35 | 15.4 | 0.97 |

## 5. Conclusion

The results from all altitude (1.5-13.5km.), the appropriate altitude range for rainfall estimation is 2.5-5.5km. Because the set of data after selected altitude range have more correlation than before selected and have less errors than before selected altitude (**Table 1.**) and the Z-R relationship obtained from the selected altitude range (2.5-5.5km.) also provide more accuracy than before selected altitude range (**Table 2.**) and when evaluate Z-R relationship of each selected altitude and calculate statistics, the appropriate altitude to improve accuracy in rainfall estimation from radar reflectivity data is 3.5 km above mean sea level. (**Table 3.** and **Table 4.**). Because the water droplet in raincloud in altitude 3.5 km is the most relate to rainfall rate measured by the rain gauge station.

## References

[1] Collinge, V. K. and Kirby, C.: Weather radar and flood forecasting, John Wiley & Sons, Great Britain, England, 1987.

[2] Robert A. Houze, Jr.: Structures of atmospheric precipitation systems: A global survey, Department of Atmospheric Sciences, University of Washington, Seattle, Washington, 1981.

[3] Chumchean, S., Aungsuratana, P., Khommuang, A. and Hanchoowong, R.: Study of Rain-Cloud Characteristics Using Weather Radar Data, Department of Civil Engineering, Mahanakorn University of Technology, Bangkok, Thailand, Bureau of Royal Rainmaking and Agricultural Aviation, Bangkok, Thailand, Department of Civil Engineering, Thammasat University, Bangkok, Thailand, 2009.

[4] Chumchean, S.: Improved estimation of radar rainfall for use in hydrological modeling, Ph.D. thesis, The University of New South Wales, Sydney, Austra1ia, 2004.

[5] Mapiam, P., Sriwongsitanon, N.: Climatological Z-R relationship for radar rainfall estimation in the upper Ping river basin, Department of Water Resources Engineering, Faculty of Engineering, Kasetsart University, Bangkok, Thailand, 2008.

[6] Piman, T., Babel, M. S., Das Gupta, A. and Weesakul, S.: Development of a window correlation matching method for improved radar rainfall estimation, Water Engineering and Management, Asian Institute of Technology, Pathumthani, Thailand, 2007.

IOP Conf. Series: Journal of Physics: Conf. Series **901** (2017) 012046 doi:10.1088/1742-6596/901/1/012046

# Investigating differences in light stable isotopes between Thai jasmine rice and Sungyod rice

**C Kukusamude and S Kongsri***

Nuclear Research and Development Division, Thailand Institute of Nuclear Technology (Public Organization), 9/9 Moo 7, Tambon Saimoon, Ongkharak, Nakhon Nayok 26120, Thailand.

*E-mail : supalak@tint.or.th

**Abstract**. We report the differences in light stable isotopes between two kinds of Thai rice (Thai jasmine and Sungyod rice). Thai jasmine rice and Sungyod rice were cultivated in the northeast and the south of Thailand. Light isotopes including 13C, 15N and 18O of Thai jasmine rice and Sungyod rice samples were carried out using isotope ratio mass spectrometry (IRMS). Thai jasmine rice (Khao Dawk Mali 105) was cultivated from Thung Kula Rong Hai area, whereas Sungyod rice was cultivated from Phathalung province. Hypothesis testing of difference of each isotope between Thai jasmine rice and Sungyod rice was also studied. The study was the feasibility test whether the light stable isotopes can be the variables to identify Thai jasmine rice and Sungyod rice. The result shows that there was difference in the isotope patterns of Thai jasmine rice and Sungyod rice. Our results may provide the useful information in term of stable isotope profiles of Thai rice.

## 1. Introduction

The most important crop in the world is rice, *Oryza sativa L.*, providing up to 76% for the population of Southeast Asia [1]. Rice is the main source of calories and protein for human especially in Asia, America, and some European countries. It also a good source of energy, vitamins, and essential elements for human [2]. Thai jasmine rice (Khao Dawk Mali 105) is originated from Thailand, mostly grown in Thung Kula Rong Hai area in the northeast of Thailand [3]. Thung Kula Rong Hai area covers five provinces including Roi-Et, Mahasarakham, Sisaket, Yasothon, and Surin. It is considered as premium rice in the world market leading to economic adulteration problem. Sungyod is mostly grown in Phatthalung province, Thailand. Sungyod is recognized as Geographical Indication (GI) rice. Thus, tracing the geographical origin of rice is essential to prevent mislabeling and adulteration [3,4,5]. Nowadays, identification of the geographical origin of a wide range of agricultural food products is interested issue. Tracing of food authenticity is important for quality control and food safety.

Various analytical techniques for discrimination of rice origin such as inductively coupled plasma atomic emission spectrometry (ICP-AES), inductively coupled plasma mass spectrometry (ICP-MS), and elemental analyzer/ isotope ratio mass spectrometry (EA/IRMS) [2,3,4,6] have been used. Stable isotope analyses have been widely used to trace the origin of organic materials in various fields such as geochemistry, biochemistry, archaeology, and petroleum chemistry [7,8]. It has become important issue as a solution for food authenticity problems. In fact, the isotopic compositions of plant reflect various factors. For instance, carbon isotopic composition of plant strongly depends on the carbon fixation process such as $C_3$ or $C_4$ cycle [9]. Nitrogen isotope mainly depends on soil nutrition [10].

In this study, we determined carbon, nitrogen, and oxygen isotopic compositions ($\delta^{13}C$, $\delta^{15}N$, $\delta^{18}O$) of Thai jasmine rice (Khao Dawk Mali 105) and Sungyod rice cultivated in the Thung Kula Rong Hai area and Phatthalung province, respectively using elemental analyzer isotope ratio mass spectrometry (EA-IRMS). In order to investigate the differences in stable isotopic compositions that can be used as parameters for discrimination of the rice origin, t-test was also evaluated.

## 2. Experimental

### *2.1. Rice samples*

Samples of two kinds of Thai rice were collected from paddy fields from different regions (the Northeast and the South) in Thailand. The samples (n=49) named Khao Dawk Mali 105 from the Northeast located in Thung Kula Rong Hai area were collected. The samples (n=43) of Sungyod rice were collected from Phatthalung province in the South of Thailand. Samples were dried and ground to fine powder before analysis.

### *2.2. $\delta^{13}C$, $\delta^{15}N$, and $\delta^{18}O$ analyses by Elemental analyzer/ isotope ratio mass spectrometry*

All samples were analyzed using elemental analyzer isotope ratio mass spectrometry (EA-IRMS), Isoprime, UK in order to determine $\delta^{13}C$, $\delta^{15}N$, *and* $\delta^{18}O$ isotope ratios. The powered rice samples were accurately weighed 4 mg into tin boats (4 mm x 4 mm x 11 mm). The boats were then folded and compressed in order to avoid air presented. For the analysis of stable nitrogen, the powered rice samples were (4 mg) was weighed and analyzed in a similar way as the stable carbon analysis. The samples for $\delta^{13}C$ and $\delta^{15}N$ analyses were performed in combustion mode by EA-IRMS. For $\delta^{18}O$ analysis, the powered rice samples (0.3 mg) were accurately weighed into silver boats (4 mm x 4 mm x 11 mm) and analyzed using pyrolysis mode.

## 3. Results and discussion

The isotope ratio was described in term of δ (per mil, ‰) defined as the following equation: $\delta$ (‰) = $(R_{sample} - R_{standard})/R_{standard}$ x 1000, where δ (‰) is the isotope composition, Rsample and Rstandard are the isotope ratio (i.e., $\delta^{13}C/\delta^{12}C$, $\delta^{15}N/\delta^{14}N$, and $\delta^{18}C/\delta^{16}C$) of the sample and isotope ratio of international standard, respectively. In order to obtain the differences in stable isotopes according to the geographical origins, rice samples cultivated from Thung Kula Rong Hai area and Phatthalung province were collected. Thung Kula Rong Hai covers 5 provinces including Sisaket, Yasothon, Roi Et, Surin, and Mahasarakham provices. The carbon, nitrogen, and oxygen isotopic compositions of rice were analyzed by EA-IRMS.

The results of $\delta^{13}C$, $\delta^{15}N$, and $\delta^{18}O$ values of Thai jasmine and Sungyod rice obtained from Thung Kula Rong Hai area and Phatthalung province are summarized in Fig. 1. Box plots A and D represent $\delta^{13}C$ values Thai jasmine rice from Thung Kula Rong Hai area and Sungyod rice from Phatthalung provinces, respectively, whereas box plots (B and E) and (C and F) represent $\delta^{15}N$, and $\delta^{18}O$ values, correspondingly. The $\delta^{13}C$ values of Thai jasmine rice ranged from -28.04‰ to -26.31‰ (n=49), while those of Sungyod rice ranged from -28.41‰ to -26.81‰ (n=43). It is found that the $\delta^{13}C$ values of all rice samples were in the range of $\delta^{13}C$ values of C-3 plants (-22‰ to -33‰). The $\delta^{15}N$ values of Thai jasmine rice were between +1.50‰ to +6.47‰. The $\delta^{15}N$ values found in Sungyod rice ranged from +3.06‰ to +5.97‰. The S.D. of those values was 1.07 and 0.64, respectively. From Fig. 2 (B, C, E, and F), the results showed that data were highly deviated compared with the $\delta^{13}C$ values. This is probably because rice samples were collected from the paddy fields used both organic and inorganic fertilizers. Moreover, Thai jasmine rice and Sungyod rice were cultivated from different regions of Thailand. The $\delta^{18}O$ values in Thai jasmine rice and Sungyod rice ranged from +23.44‰ to +26.69‰ and +23.50‰ to +29.50‰, respectively. The heaviest was obtained in Sungyod rice sample cultivated in Phatthalung province whereas the lightest was the Thai jasmine rice sample cultivated in Thung Kula Rong Hai area.

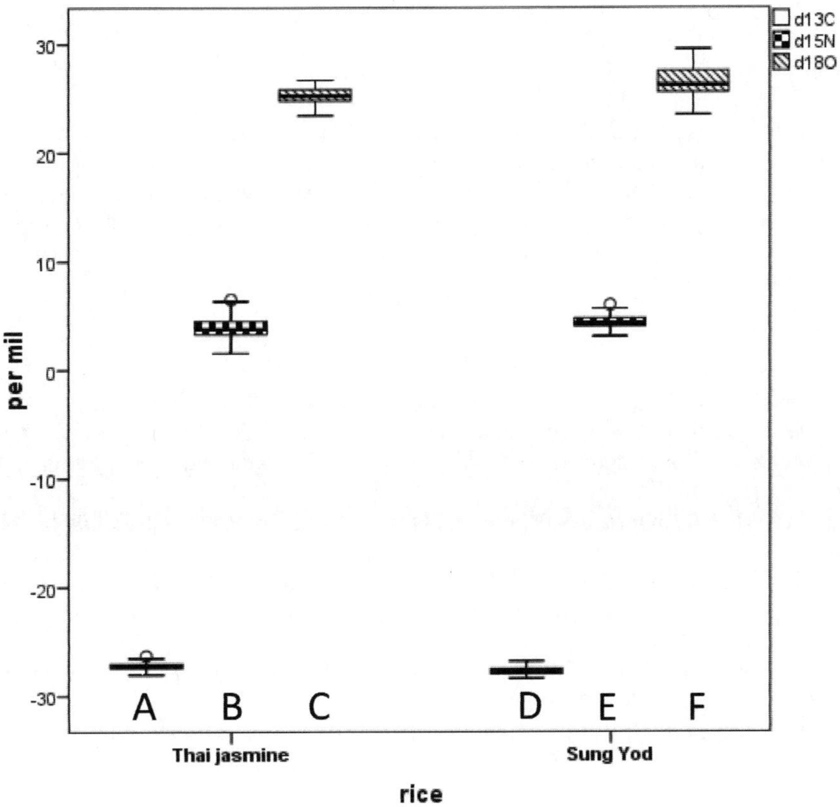


**Figure 1.** Box plot of $\delta^{13}C$, $\delta^{15}N$, and $\delta^{18}O$ values of rice originated from Thung Kula Rong Hai area and Phatthalung province.

**Table 1.** Descriptive statistic and t-test for Thai jasmine rice and Sungyod rice.

| Isotope | Average (Thai jasmine rice) | Average (Sungyod rice) | S.D. (Thai jasmine rice) | S.D. (Sungyod rice) | P (T<=t) |
|---|---|---|---|---|---|
| $\delta^{13}C$ | -27.24 | -27.71 | 0.32 | 0.37 | <0.0001 |
| $\delta^{15}N$ | +3.93 | +4.34 | 1.07 | 0.64 | 0.027 |
| $\delta^{18}O$ | +25.28 | +26.36 | 0.84 | 1.39 | <0.0001 |

The average and S.D. of the interested elements found in Thai jasmine rice and Sungyod rice cultivated in Thung Kula Rong Hai area and Phatthalung province are illustrated in the Table 1. The difference in isotopic composition between Thai jasmine rice and Sungyod rice was carried out by t-test. The obtained p-values of $\delta^{13}C$, $\delta^{15}N$, and $\delta^{18}O$ were <0.0001, 0.027 and <0.0001, respectively. It indicated that all studied isotopes including $\delta^{13}C$, $\delta^{15}N$, and $\delta^{18}O$ were significantly different between Thai jasmine rice and Sungyod rice samples at 95% confidence interval. This is possibly due to there was different in the geographical origin and water resource that used for cultivation of rice. This implied that the carbon, nitrogen, and oxygen isotopes could be possible to identify rice samples cultivated Thung Kula Rong Hai area (in the northeastern region of Thailand) and Phatthalung province. Our results may be useful database of Thai jasmine rice and Sungyod rice and may provide preliminary information for discrimination of rice samples that grown in different regions in Thailand.

## 4. Conclusion

Three isotopes ($\delta^{13}C$, $\delta^{15}N$, and $\delta^{18}O$) in Thai jasmine and Sungyod rice samples cultivated from Thung Kula Rong Hai area and Phatthalung province were determined by EA-IRMS. All studied isotopes were significantly different between Thai jasmine rice and Sungyod rice samples at 95% confidence interval. Our results may provide preliminary information for discrimination of rice samples and it is possible to trace the geographical origin of rice cultivated from different regions in the further study using stable isotopes.

## References

[1] Fitzgerald M A, McCouch S R and Hall R D 2009 *Trends Plant Sci.* **14** 133.
[2] Promchan J, Günther D, Siripinyanond A and Shiowatana J 2016 *J. Cereal Sci.* **71** 198.
[3] Cheajesadagul P, Arnaudguilhem C, Shiowatana J, Siripinyanond A and Szpunar J 2013 *Food Chem.* **141** 3504.
[4] Kelly S, Baxter M, Chapman S, Rhodes C, Dennis J and Brereton P 2002 *Eur Food Res Technol* **214** 72.
[5] Vemireddy L R, Satyavathi V V, Siddiq E A and Nagaraju J 2015 *J. Food Sci. Tech.* **52** 3187.
[6] Suzuki Y, Chikaraishi Y, Ogawa N O, Ohkouchi N and Korenaga T 2008 *Food Chem.* **109** 470.
[7] Ambrose S H, DeNiro M J 1986 *Nature* **319** 59.
[8] Hayes J M, Freeman K H, Popp B N and Hoham C H 1990 *Org. Geochem.* **16** 1115.
[9] Smith B N and Epstein S 1971 *Plant Physiol.* **47** 380.
[10] Meints V W, Shearer, G, Kohl, D H and Kurtz L T 1975 *Soil Sci.* **119** 421.

**Acknowledgements**

Financial support from Thailand Institute of Nuclear Technology is gratefully acknowledged.

# A preliminary study on identification of Thai rice samples by INAA and statistical analysis

**S Kongsri and C Kukusamude**

Nuclear Research and Development Division, Thailand Institute of Nuclear Technology (Public Organization), 9/9 Moo 7, Tambon Saimoon, Ongkharak, Nakhon Nayok 26120, Thailand.

*E-mail : chunyapuk@tint.or.th

**Abstract**. This study aims to investigate the elemental compositions in 93 Thai rice samples using instrumental neutron activation analysis (INAA) and to identify rice according to their types and rice cultivars using statistical analysis. As, Mg, Cl, Al, Br, Mn, K, Rb and Zn in Thai jasmine rice and Sung Yod rice samples were successfully determined by INAA. The accuracy and precision of the INAA method were verified by SRM 1568a Rice Flour. All elements were found to be in a good agreement with the certified values. The precisions in term of %RSD were lower than 7%. The LODs were obtained in range of 0.01 to 29 mg $kg^{-1}$. The concentration of 9 elements distributed in Thai rice samples was evaluated and used as chemical indicators to identify the type of rice samples. The result found that Mg, Cl, As, Br, Mn, K, Rb, and Zn concentrations in Thai jasmine rice samples are significantly different but there was no evidence that Al is significantly different from concentration in Sung Yod rice samples at 95% confidence interval. Our results may provide preliminary information for discrimination of rice samples and may be useful database of Thai rice.

## 1. Introduction

Thai Jasmine Rice (Khao Dawk Mali 105) is one of the most important consuming products in Thailand. Khao Dawk Mali 105 is originated from Thailand, mostly grown in Thung Kula Rong Hai area in the northeast of Thailand [1]. Thung Kula Rong Hai area covers five provinces including Roi-Et, Mahasarakham, Sisaket, Yasothon, and Surin. It is not only for the local consumption but also the one of the main export products to all over the world. Thai jasmine rice cultivated in Thung Kula Rong Hai area in the northeastern region is the most famous premium rice which has a unique sweet-scented aroma, softness, and tender texture [2]. Sungyod rice is originated from Phatthalung province and recognized as Geographical Indication (GI) rice. Therefore, adulteration of Thai Jasmine Rice and Sungyod rice with other cultivation area leads to rice authenticity problems by mislabeled addition of inferior rice to premium rice. However, the geographical origin of rice cannot be distinguished by image analysis [3]. Thus, it is important to resolve these rice authenticity problems. Recently, elemental compositions have been used as a chemical indicator for discrimination of cultivation areas [4].

Several analytical techniques that have been published for the detection of elemental compositions are inductively coupled plasma optical emission spectrometry (ICP-OES) [5], inductively coupled plasma mass spectrometry (ICP-MS) [1], flame atomic absorption spectroscopy (FASS), graphite furnace atomic absorption spectrophotometry (GF-AAS) [6] and so on. However, these methods require many time-consuming sample preparation steps prior to analysis. Moreover, the risk of samples contamination and loss of analytes can be occurred in the sample preparation procedures. Neutron activation analysis (NAA) is non-destructive method [7] suiting for solid samples that are difficult to dissolve. The sample preparation for NAA is not necessary, thus it can reduce the risk of contamination and analysis time [8].

In this work, we illustrated the preliminary study on identification the origin of Thai jasmine rice and Sungyod rice cultivated from different regions of Thailand, Thung Kula Rong Hai area and Phatthalung province using INAA and statistics. The quantification based on elemental composition by INAA was determined by comparative method. Certified reference material was also checked.

## 2. Experimental

### *2.1. Rice samples*

Samples of two kinds of Thai rice were collected from paddy fields from different regions (the Northeast and the South) in Thailand. The samples (n=45) of Thai jasmine rice (Khao Dawk Mali 105) from Thung Kula Rong Hai area in the Northeast of Thailand were collected. The samples (n=48) of Sungyod rice were collected from Phatthalung province in the South of Thailand. Samples were dried and pulverized to fine powder by agate mortar. Then, the samples were dried at 60 °C for 48 h in oven and kept in desiccator prior to analysis.

### *2.2. Instrumental neutron activation analysis*

The samples and reference material (approximately 100 mg) were weighed and sealed in polyethylene bags prior to neutron irradiation. These containers were packed into rabbit for irradiation. The irradiation condition and other details of the elemental analysis are presented in Table 1. After the appropriate decay time, the gamma-ray activities of the radioisotopes were counted using a high purity germanium HPGe detector operated by the Gamma Vision computer program (EG & ORTEC, USA). In this study, Magnesium (Mg), Aluminium (Al), Chlorine (Cl), Arsenic (As), Bromine (Br), Manganese (Mn), (K), Rubidium (Rb), and Zinc (Zn) in rice samples were quantified by INAA.

**Table 1** The irradiation conditions for the analysis of elements in Thai rice samples

| Irradiation condition | Irradiation Facility | Irradiation time | Cooling time | Counting time | Radionuclide |
|---|---|---|---|---|---|
| Short life radionuclide | Pneumatic system | 25 s | 2 m | 300 s | Mg Al Cl |
| Medium life radionuclide | CA-2 | 7 h | 12 h | 1800 s | As Br Mn K |
| long life radionuclide | Lazy Susan | 26 h | 14 d | 3600 s | Rb Zn |

## 3. Results and discussion

In order to identify Thai rice samples according to the geographical origins, rice samples cultivated from Thung Kula Rong Hai area and Phatthalung province were collected. Thung Kula Rong Hai covers 5 provinces including Sisaket, Yasothon, Roi Et, Surin, and Mahasarakham provices. The multi-elemental compositions of rice were analyzed by INAA. The accuracy and precision of the INAA method were verified by SRM 1568a Rice Flour. All concentration values were found to be in a good agreement with the certified values. The precisions in term of %RSD were lower than 7%. The LODs were obtained in range of 0.01 to 29 mg $kg^{-1}$.

The results of Mg, K, As, Br, Cl, Al, Mn, Rb, and Zn concentration of Thai jasmine and Sungyod rice obtained from Thung Kula Rong Hai area and Phatthalung province are summarized in Fig. 1. Fig. 1A shows box plots of Mg and K concentrations in Thai jasmine rice and Sungyod rice. Mg found in Thai jasmine rice and Sungyod rice were in the ranges of 0.048-0.083 and 0.034-0.088 %(w/w), respectively, while K was found in the ranges of 0.089-0.178 and 0.072-0.130 %(w/w), correspondingly. As and Br (Fig. 1B) found in Thai jasmine rice were 0.043-0.405 ppm and 0.015-0.302 ppm and those in Sungyod rice were 0.067-0.400 ppm and 0.080-0.403 ppm, respectively. Cl found in Thai jasmine rice and Sungyod rice were in the ranges of 149.476-617.284 and 99.129-316.656 ppm, respectively, while Al was found in the ranges of 76.375-122.785 and 71.946-125.526 ppm, correspondingly, as shown in Fig. 1C. Mn, Rb, and Zn in Thai jasmine rice and Sungyod rice are shown in Fig. 1D. Mn and Rb found in Thai jasmine rice were 8.222-21.450 ppm and 2.077-10.080 ppm and those in Sungyod rice were 7.822-16.215 ppm and 6.197-50.544 ppm, respectively. Zn found in the ranges of 13.930-28.870 and 7.971-28.524 ppm for Thai jasmine rice and Sungyod rice samples, respectively.

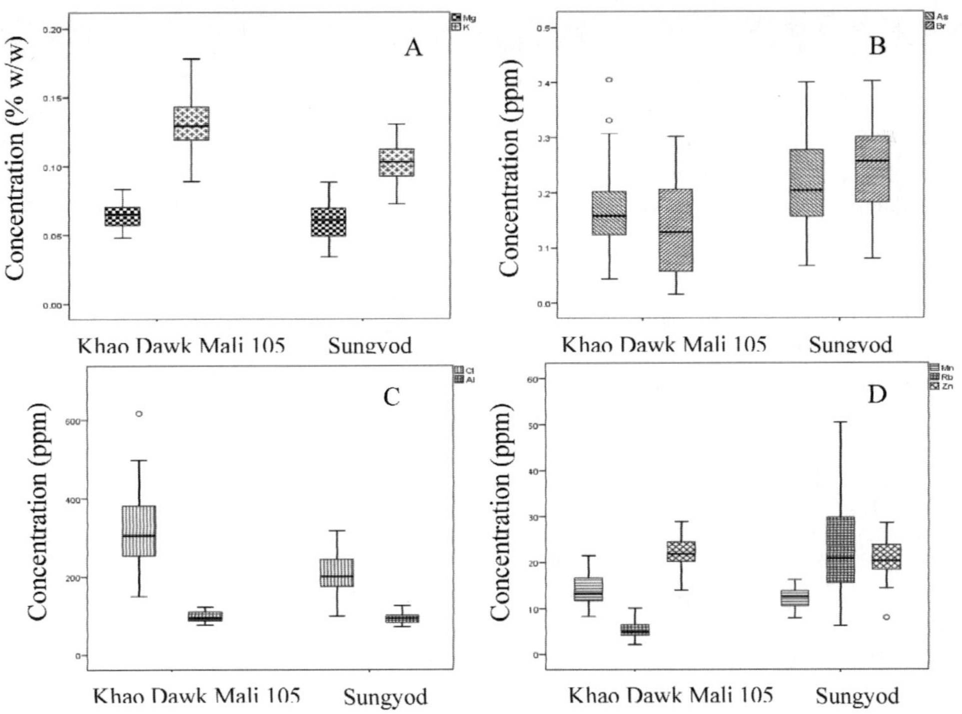


Fig. 1 Box plots of the concentration of 9 elements in Thai jasmine rice and Sungyod rice samples

The average and RSD of the interested elements found in Thai jasmine rice and Sungyod rice are illustrated in the Table 2. T-test was evaluated for testing the difference of elemental composition between Thai jasmine rice and Sungyod rice grown in different regions of Thailand. The p-values of Mg, Cl, Al, As, Br, Mn, K, Rb, and Zn were 0.043, <0.0001, 0.067, 0.003, <0.0001, 0.007, <0.0001, <0.0001, and 0.045, respectively. It indicates that Mg, Cl, As, Br, Mn, K, Rb and Zn concentrations were significantly different but there was no significant difference in Al concentration between Thai jasmine rice and Sungyod rice samples at 95% confidence interval. This was implied that the elemental composition could be good indicators and possible to identify rice samples.

**Table 2** Descriptive statistical analysis of elements in Thai jasmine rice and Sungyod rice samples.

| Element | Average (Thai jasmine) | Average (Sung Yod ) | RSD (Thai jasmine) | RSD (Sung Yod) | P(T≤Two-tail) |
|---|---|---|---|---|---|
| Mg(%) | 0.065 | 0.060 | 0.154 | 0.217 | 0.043 |
| Cl (mg $kg^{-1}$) | 319.696 | 205.783 | 0.295 | 0.244 | <0.0001 |
| Al (mg $kg^{-1}$) | 97.472 | 92.899 | 0.124 | 0.126 | 0.067 |
| As (mg $kg^{-1}$) | 0.166 | 0.217 | 0.476 | 0.364 | 0.003 |
| Br (mg $kg^{-1}$) | 0.132 | 0.246 | 0.606 | 0.337 | <0.0001 |
| Mn(mg $kg^{-1}$) | 13.880 | 12.300 | 0.228 | 0.175 | 0.007 |
| K (%) | 0.131 | 0.102 | 0.153 | 0.137 | <0.0001 |
| Rb (mg $kg^{-1}$) | 5.406 | 22.794 | 0.374 | 0.443 | <0.0001 |
| Zn (mg $kg^{-1}$) | 22.057 | 20.541 | 0.152 | 0.185 | 0.045 |

## 4. Conclusion

Nine variables including Mg, Cl, Al, As, Br, Mn, K, Rb, and Zn in Thai jasmine and Sungyod rice samples cultivated from Thung Kula Rong Hai area and Phatthalung province were determined by INAA. Mg, Cl, Al, As, Br, Mn, K, Rb, and Zn found were significantly different between Thai jasmine rice and Sungyod rice samples at 95% confidence interval. There was no significant difference in Al concentration found between Thai jasmine rice and Sungyod rice samples at 95% confidence interval. It may be possible to trace the geographical origin of rice cultivated from different regions in Thailand for the further study. Our results provide preliminary information for Thai rice samples according to rice origins.

## References

[1] Cheajesadagul P, Arnaudguilhem C, Shiowatana J, Siripinyanond A and Szpunar J 2013 *Food Chem.* **141** 3504.
[2] Leelayuthsoontorn P and Thipayarat A 2006 *Food Chem.* **96** 606.
[3] Suzuki Y, Chikaraishi Y, Ogawa N O, Ohkouchi N and Korenaga T 2008 *Food Chem.* **109** 470.
[4] Kelly S, Baxter M, Chapman S, Rhodes C, Dennis J and Brereton P 2002 *Eur Food Res Technol* **214** 72.
[5] Oliveira A, Baccana N and Cadore S 2012 *J. Braz. Chem. Soc.* **23** 838.
[6] Parengam M, Judprasong K, Songsak S, Jittinandana S, Laoharojanaphand S and Busamongkol A 2010 *J Food Compost Anal.* **23** 340.
[7] Kongsri S, Srinuttrakul W, Sola P and Busamongkol A 2016 *Energy Procedia* **89** 261.
[8] Carmen C, Duliu O G, Ricman C, Toma M, Dragolici F, Bragea M and Done L 2008 *Rom. J. Phys.* **53** 941.

**Acknowledgements**

Financial support from Thailand Institute of Nuclear Technology is gratefully acknowledged.

# Potential Increasing of Rubber Sheet Production with Fungus Displacement by Solar Tunnel Dryer by Integrating a Biomass Gasifier as an Assisting Heat Source

**D Dangphonthong[1], P Ruenruangrit[1], and W Pinate[2]**

[1]Program of Engineering Management, Faculty of Engineering, Rajabhat MahaSarakham University 80, Nakhonsawan road, Talad sub-district, Mueang district, Mahasarakham province, 44000, Thailand
Email:d.dungphontong@gmail.com
[2]Program of Physics, Faculty of Science and Technology, Rajabhat MahaSarakham University 80, Nakhonsawan road, Talad sub-district, Mueang district, Mahasarakham province, 44000, Thailand, Email: kaapplied@hotmail.com

**Abstract.** In this study, Potential increasing of rubber sheet production with fungus displacement by solar tunnel dryer. By integrating a biomass gasifier as an assisting heat source for drying Rubber Sheet produce and its performance analysis has been investigated. An experimental study has been carried out with the stand-alone solar tunnel dryer and the dryer with the assisted biomass gasifier. For thermal performance analysis, for wind speed 1.0 m/s, the dryer with assisted gasifier gave an average efficiency of 32.86% higher and the drying period was 31.45 % shorter than those of the unit without the gasifier. For wind speed 1.5 m/s, and for wind speed 2.0 m/s, the values are 21.49 % , 28.68 % and 33.17% , 24.54 % , respectively. From the economic analysis, drying of wind speed 1.5 m/s is the most appropriated. The payback and the IRR were 1.8 years and 55.48 %, respectively. While those for wind speed 1.0 m/s were 2.74 years and 36%, respectively.
Keyword: Tunnel Dryer, Solar Dryer

## 1. Introduction

In dealing rubber sheets, the price is dependent on the quality of the rubber sheet. By which the quality will be tested for various values which are the amount of dirt within the rubber sheet, the amount of mold, the Inflatable point, thickness, weight, strength and flexibility, Contamination, Sand gravel and other flaws of a rubber sheet. Taken and adjusted from standards of "The Green Book" [1], one of the problems of rubber sheets found is that fungus around the sheet by which is the result from humidification non-standard drying of rubber sheets. Also, including drying rubber sheets during the raining season at a random pattern, which causes damp mildew, by which molds are created from rubber not being dry enough. If left out for humidity to be absorbed, even if not up to 1%, this can result in molding. Rubber with excessive molding will have a decrease in weight by 2% within a month, which affects the general price of the rubber. Therefore, the research team has the idea to increase the production capacity of rubber sheets by using solar energy in the drying process, to prevent molding around the rubber. This in a way increases the value of the rubber sheet.

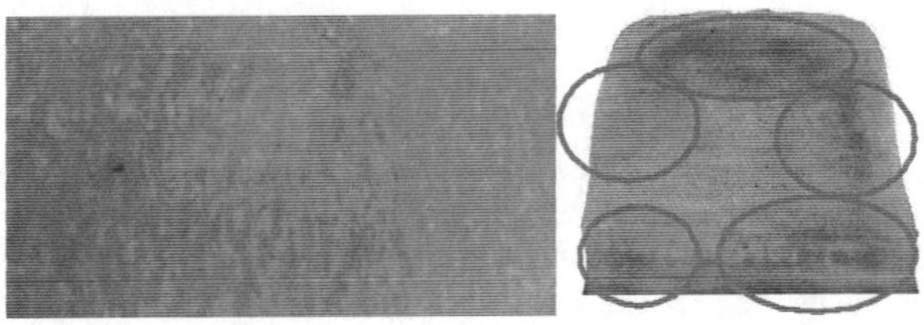

**Figure 1.** Fungus (white mold and rust mold) within the rubber sheet

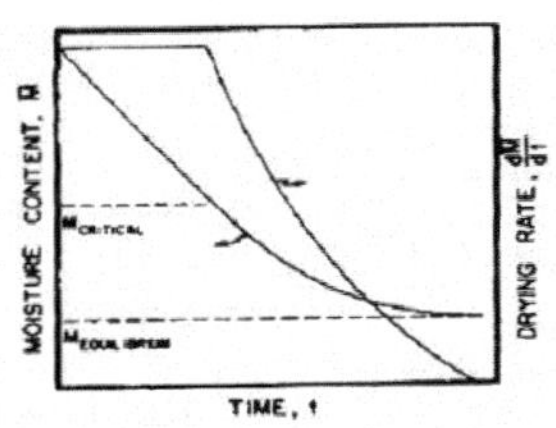


**Figure 2.** Drying at constant and reducing drying rate

To bring in solar energy to increase the quality of rubber for the community, which drying is a long-known agricultural process and there are many drying technologies in the present such as drying with hot air, solar powered drying (sun drying), vacuum drying, etc. For tunnel solar drying, there is a tunnel which has a transparent plastic sheet on top for the sunlight to penetrate through. As shown in figure 2, inside is a sieve tray placed as a line to pack the agriculture production that needs to be dried by receiving direct sunlight. The production will rise in temperature and the moisture in the production will be cast out slowly, which is sometimes the air may cause heat through the flow of the solar collector. [2-4]

## 2. Theory

### *2.1 drying*

In the process which heat is extruded with a certain method, to remove moisture by evaporation by relying on heat as a latent heat of evaporation, it is important in the process of drying to release heat into the drying material efficiently [5]. By which drying often uses hot air an intermediary in the drying process. Heat will be transplanted into the surface of the material by which most is used for evaporation. Steam will move from the surface of the material to the air flow. If the surface of the material has a large amount of liquid, the temperature and the concentration of the surface vapor will be fixed. This results in a heat transfer rate as well as a constant drying rate. If the temperature, moisture, and speed of the air flow are constant, then when the surface of the material has a decreased amount of liquid the temperature and the concentration of the steam on the surface of the material inevitably changes. By which the temperature of the material will rise and the concentration of the steam will decrease causing a decrease in the drying rate. Therefore, during the range of the drying rate the moisture between fixed drying rate range and the lower drying rate range is called "humidity", which is shown in figure 2. The majority of materials that are to be baked have an interior structure of a porous. The drying ranges can be divided into two ranges which are a fixed drying rate and a decreasing drying rate [6]. Moisture is the indicator of the amount of water contained in the material by comparing with the mass of the material that is damp or dry as the following equation [7-9].

Wet standard humidity

$$M_W = \frac{(w - d)}{w} \tag{1}$$

And dry moisture standard found from

$$M_d = \frac{(w - d)}{d} \tag{2}$$

When w is the mass of damp material (kg)
D is dry mass of material (kg)
$M_d$ is the dry moisture standard (standard dry fraction)
$M_w$ is the wet humidity standard (standard wet fraction)

### *2.2 Gasifier system*

Principles of biodegradable gasification furnace. The first step is to use charcoal as fuel. Because it can produce gas. Stable than firewood. Charcoal, used as a fuel, weighs about 1 kg, is put into a gasifier. At the same time, the charcoal is used as fuel to ignite outside the furnace. When the charcoal ignited, it poured into the charcoal before. Turn on the exhaust fan from the outside into the burner. Using a low flow rate first. Then add the charcoal into the oven for about 3 kilograms. Climb the top of the furnace to produce gas. Over time, about 20 minutes can accelerate the flow of air into the gasification furnace. By checking the air entering the gasifier furnace to see if it needs to be changed. By firing at the cyclone exit. To burn a Gas Producer that ignited it? In ignition of the cyclone. It depends on the rate of air flow into the gasification furnace is important. This would require techniques to control the flow of air. And other factors include airiness or burning pattern inside a gas stove, siphon etc. And when the Producer Gas burns well, it can be used to blow the hot gas into the tunnel solar drying system. At the same time, the testers always have to constantly check the amount of charcoal burned in the combustion chamber. And come to the car to refill charcoal into the oven as needed and in the test will be used firewood for comparison.

IOP Conf. Series: Journal of Physics: Conf. Series **901** (2017) 012050 doi:10.1088/1742-6596/901/1/012050

**Figure 3.** Biomass gasifier in this research.

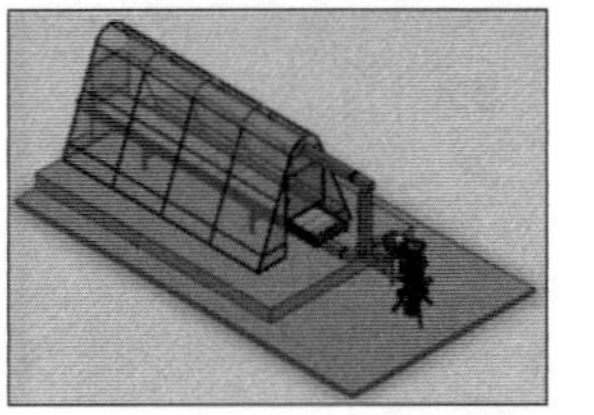

**Figure 4.** Tunnel Solar Drying Room by using biodegradable gas stove for additional heat for drying rubber sheets

### 3. Procedure
Researchers have designed a solar tunnel dryer, which has a higher tunnel. This allows people access which has a height of 2.10 meter and the base has a width of 3.20 meters as shown in figure 3. This enables the drying of rubber sheet conveniently. At the end of the tunnel, there will be a chimney to help in the drainage of moisture from the rubber sheets. The work will control the temperature by which there is a biodegradable gas stove which biomass will come from agricultural residues such as husks, branches, leaves that are gas producers. Hot air is supplied to the drying chamber at the bottom of the conveyor belt making the rubber sheet dry faster. All of this, the reason for designing the tunnel to have a biased bridge to help in moving rubber sheets from the front of the tunnel to the back with an electric motor. The speed of the biased bridge and the distance must be studied to be suitable with the size of the rubber sheets and the amount that needs to be dried.

### 4. Results
The effect of the reduction of moisture ratio of the drying process is under the frequency of belting the bridge (Hz). 3 levels of frequency include5.1 , 2.5 and 3.5 Hz and use 3 example groups of rubber sheets and compared them to 2 drying patterns which are Drying System that uses only sunlight, and a drying system that uses sunlight together with gasifier that is used to give additional heat. You will see that the temperature has an effect towards these two drying patterns by which the drying system chases out moisture that uses only sunlight. The average temperature within the drying tunnel is at about 50-55 degrees Celsius. As for the drying system that chases out moisture that uses sunlight together with gasifier which uses additional heat will have an average temperature of about 65-70 degrees Celsius. The high drying temperature will make the ratio of moisture have a fast decrease which is a result that causes the drying rate to be high. Also, it takes less time in the drying process when considering the humidity ratio of each temperature. It is found that each drying temperature, in the starting period, the moisture ratio will decrease speedily and also relatively slow down until once time passes by, in the drying process it is found that the moisture ratio tends to decrease steadily. Since in the beginning of drying all three groups of the rubber sheets, there is still high moisture which the heat in the drying tunnel has the ability to make the amount of liquid evaporate out of the production well. Then makes the moisture ratio, in the beginning, decrease fast and when able to dry up to a certain stage, the amount of liquid in the production starts to decrease in amount. This makes the moisture ratio start to slowly decrease when a time of the drying process passes by, making the moisture ratio almost fixed. From the analysis of the both drying systems, it is also found that the drying process that chases out moisture with sunlight together with gasifier that uses additional heat has a high-speed drying ratio, more than the drying system that chases out moisture by using sunlight alone.

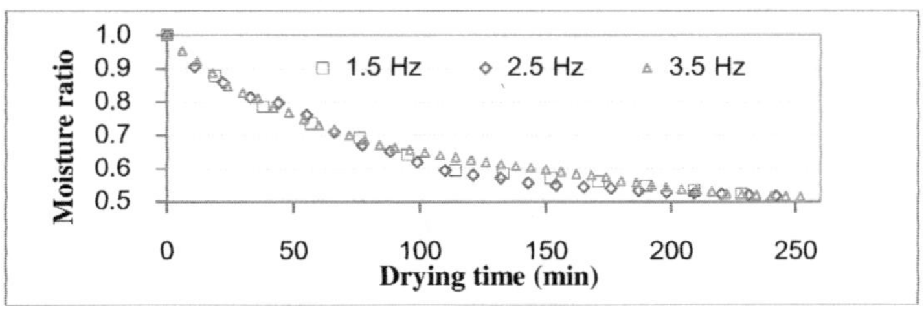


**Figure 5.** Comparing ratio of moisture of rubber sheets in example group 1 with time of drying system that uses sunlight and gasifier

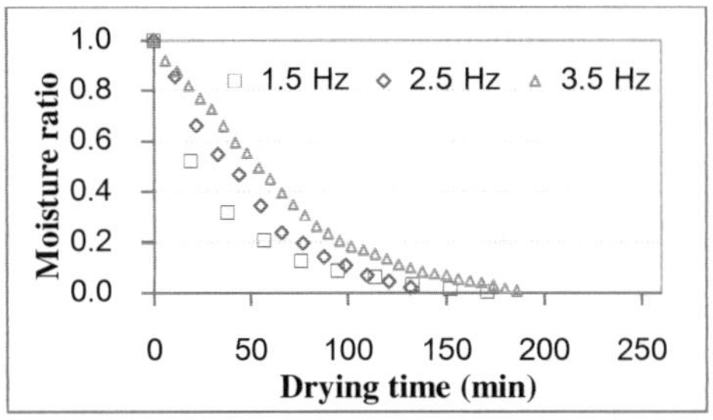


**Figure 6.** Comparing ratio of moisture of rubber sheets in example group 2 with time of drying system that uses sunlight and gasifier

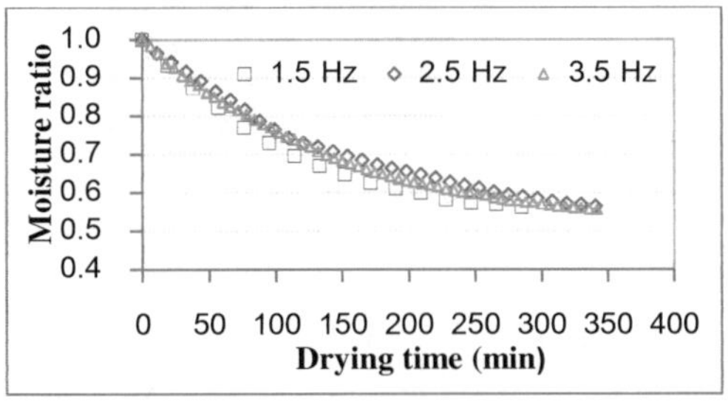


**Figure 7.** Comparing ratio of moisture of rubber sheets in example group 3 with time of drying system that uses sunlight and gasifier

## 5. Conclusion

*5.1 Studying the drying system of rubber sheets of example group 1*

The system using sunlight along with gasifier that gives additional heat in efficient drying has efficiency of drying higher than average of about 32.86 % and dries in a time below the average of about 31.45 % (if the system that uses sunlight in drying in 100 hours, the system using sunlight with gasifier will take about 68.67 hours).

*5.2 Studying the drying system of rubber sheets of example group 2*

The system using sunlight with gasifier that gives additional heat in drying has the efficiency of drying higher than the average of about 21.49 % and takes time in drying lower than the average of about 28.68 % (if the system that uses sunlight takes 100 hours, the system that uses sunlight along with gasifier will take about 71.31 hours).

*5.3 Studying the drying system of rubber sheets of example group 3*

The system using sunlight with gasifier that gives additional heat in drying has the efficiency of drying higher than the average of about 33.17 % and takes time in drying lower than the average of about 24.54 % (if the system that uses sunlight takes 100 hours, the system that uses sunlight along with gasifier will take about 75.45 hours).

## Acknowledgements

Thank you to the National Research Council of Thailand (NESDB) and Rajabhat MahaSarakham University for supporting the research budget of the research team and helping in checking bug fixes in the working with care and attention from start to finish.

## References

[1] Senadeera W 2004 *Thesis of the Department of Agriculture* The Ministry of Agriculture and Cooperatives Technical Paper pp 97-104
[2] Gauhar A, Mastekbayeva M and Kumar S 1998 *ASEAN Seminar and Workshop on Drying Technology* 3-5 June Phitsanulok Thailand pp 119-136
[3] Gauhar A, Chandika M, Augustus M, And Kumar S 1999 *ISES 99 Solar World Congress*, Israel, July, p 346
[4] Bonggoch P and Sukreudee N 2007 *The 3rd Thailand Energy Network Conference* pp139- 144
[5] Kittikorn S, Weera F, Pawin M, and D. Natawut 2007 *Conference on Heat and Mass Transfer in Thermal Devices 6th.* Amora Hotel Chiang Mai 223 - 216
[6] Haquel M And Langrish T 2005 *J. Dry. Tech.* 3(23) pp 1541–1553
[7] Gauhar A, Mastekbayeva M and Kumar S 1998 *Proceedings of ASEAN Seminar and Workshop on Drying Technology* Phitsanulok Thailand pp 76-88
[8] Gauhar A, Chandika P, Augustus M And Kumar S 1999 *Proceedings of ISES 99 Solar World Congress* Israel pp 226-235
[9] Jain D and Tiwari G 2000 *J.Bio. Tech.* 2(52) pp 283 – 285.

# Atmospheric water vapor: Distribution and Empirical estimation in the atmosphere of Thailand

**S Phokate**

Applied Physics, Faculty of Engineering, Khon Kaen Campus, Rajamangala University of Technology Isan, Khon Kaen, Thailand

**E-mail:** syphokate@hotmail.com

**Abstract.** Atmospheric water vapor is a crucial component of the Earth's atmosphere, which is shown by precipitable water vapor. It is calculated from the upper air data. In Thailand, the data were collected from four measuring stations located in Chiang Mai, Ubon Ratchathani, Bangkok, and Songkhla during the years 1998-2013. The precipitable water vapor obtained from this investigation were used to define an empirical model associated with the vapor pressure, which is a surface data at the same stations. The result shows that the relationship has a relatively high level of reliability. The precipitable water vapor obtained from the upper air data is nearly equal to the value from the model. The model was used to calculate the precipitable water vapor from the surface data 85 stations across the country. The result shows that seasonal change of the precipitable water vapor was low in the dry season (November-April) and high in the rainy season (May-October). In addition, precipitable water vapor varies along the latitudes of the stations. The high value obtains for low latitudes, but it is low for high latitudes.

## 1. Introduction

Water vapor is a critical component of Earth's climate systems. It is the Earth's primary greenhouse gas, trapping more heat than carbon dioxide [1,2]. Knowledge of accurate water vapor quantification and interpretation of their physical characters are crucial in understanding how the global climate system such as global warming changing over time [3]. In addition, the water vapor in the atmosphere is an important influence in the solar radiation depletion that comes through the atmosphere to the Earth [4]. The water vapor can absorb more than 10% of solar radiation traveling through the atmosphere, while the amount of absorption depends on the water vapor in the atmosphere [5-7]. Generally, the quantity of water vapor in the atmosphere is shown by precipitable water vapor. Precipitable water vapor represented by the height of the water column of the atmosphere from the surface of Earth to the top of the atmosphere. Assuming that the water vapor insert in the column of the atmosphere condensation becomes water. The height of the water in the vertical is precipitable water vapor [7-9]. In Thailand, there is no direct measurement. However, it was found that water vapor quantity is closely related to the relative humidity and the temperature.

In this study, our aim is to find a mathematical model for calculating the precipitable water vapor in the atmosphere of Thailand. The findings will help to lower the cost of measure the value of water vapor for the stations which has no upper air data. The use of empirical model is a common practice. It is used as basic data in the study about incoming solar radiation in the Earth's surface. In addition, detailed data of the water content are important input data for hydrological, energetic and radiation models [2,5,10-11].

## 2. Materials and methods

The precipitable water vapor is calculated from the relation between the relative humidity and temperature of the upper air data at four meteorological stations, namely Chiang Mai (18.78 °N, 98.98 °E), Ubon Ratchathani (15.25 °N, 104.87 °E), Bangkok (13.73 °N, 100.57 °E) and Songkhla (7.20 °N, 100.60 °E) during the years 1993-2013. The precipitable water vapor from upper air data which are the observe values. It is determined using the following equation [3,8,12]:

$$w = -\frac{1}{\rho g}\int_{p_1}^{p_2} M_p dp \tag{1}$$

where $w$ is precipitable water vapor (cm), g is the acceleration due to the earth's gravity (986.665 cm $s^{-2}$), $\rho$ is the density of liquid water (g $cm^{-3}$), p is the atmospheric pressure along altitude (mbar), $p_1$ and $p_2$ are the atmospheric pressure at the Earth's surface and top of the atmosphere respectively. $M_p$ is mixing ratio at the pressure level, p. The integration is from the surface at $p_1$ up to the pressure designated by $p_2$ which depends on the final altitude reached by the radiosonde observation. $M_p$ can be calculated by [1,7].

$$M_P = \frac{0.622e}{p-e} \tag{2}$$

where $e$ is the actual vapor pressure (mbar). The actual vapor pressure is obtained as the product of the saturated vapor pressure ($e_s$) and the relative humidity ($RH$) at pressure p [4,7].

$$e = \frac{RHe_s}{100} \tag{3}$$

The saturated vapor pressure value depends only on the air temperature ($T$) in degree Celsius. The saturated vapor pressure (in mbar) is calculated according to [13-14]:

$$e_s = 6.112\exp\left(\frac{17.67T}{T+243.5}\right) \tag{4}$$

The precipitable water vapor from upper air data was analyzed to find the relation between temperature and relative humidity which is surface data from the same stations. The researcher test the performance of the proposed model with independent data from the year 2014 in order to confirm the accuracy of the model. The performance of model was assessed based on some commonly used statistical indicators. Three statistical estimators of mean bias error (MBE), mean percentage error (MPE) and root mean square error (RMSE) are used to evaluate the accuracy of the model [1,7,12].

## 3. Results and discussion

The relation between precipitable water vapor value from upper air data, air temperature and relative humidity which is surface data from the same station were calculated using the average daily during the years 1993- 2013. The data of w from the upper air data and the actual vapor pressure from surface data at the same station. The least square straight line fit to these data yields the relation to be used for an estimate of precipitable water vapor. The empirical model under investigation is as follows.

$$w_p = 1.461407\exp(0.0451e) \tag{5}$$

where $w_p$ is precipitable water vapor (cm) which is used as a model. The $e$ is the actual vapor pressure (mbar), which depends temperature and relative humidity at surface. The precipitable water vapor obtained from upper air data is nearly equal to the value of the surface data. It was found that predominantly high correlations and relatively high coefficients of variation were obtained for the linear relationships at all the stations. The MBE was 0.029 cm, the MPE was 8.196%, shows the RMSE was 0.521 cm, and the correlation coefficient ($R^2$) was 0.875.

The model is retested with independent data, which has never been used in the modeling by using the data of year 2014 of the same stations to verify the accuracy of the model. The findings are presented in the model as shown in table 1. This result was compared to that reported by this research, Okulov et al. [1], Adeyemi [3] and Leckner [9] based on the relationship between surface temperature and relative humidity, which is the same data at the same time. The actual value of the upper air data ($\overline{X}$= 4.278± 0.922 cm) were compared. Model in this research showed the best performances according to its RMSE, MBE and MPE. Shows the RMSE was 0.525 cm, the MBE was -0.128 cm and the MPB was 10.844% (as shown in table 1), which are in good agreement than other methods.

**Table 1.** $\overline{X}$, RMSE, MBE and MPE of the precipitable water vapor difference obtained from the comparison study between this researches with the researches Okulov et al., Adeyemi and Leckner.

| Method | $\overline{X}$ (cm) | RMSE (cm) | MBE (cm) | MPE (%) |
|---|---|---|---|---|
| This research | 4.406±0.785 | 0.525 | -0.128 | 10.844 |
| Oculi et al.[1] | 3.243±0.431 | 1.198 | 0.431 | 14.108 |
| Adeyemi [3] | 3.481±0.524 | 0.973 | 0.797 | 23.004 |
| Leckner [9] | 4.503±0.665 | 0.592 | -0.225 | 12.573 |

The average daily from surface data during the years 1993-2013 of 85 stations across the country were calculated by the model. Found that, precipitable water vapor is quite high in rainy season, and a decrease in the dry season. In addition, water vapor has changed the location and month of the year. Consistent with Okulov et al. [1], Smirnov and Moore [10] and Maghrabi and Clay [12], finds that the amount of water vapor depends on the surface data and the seasonal changes. The experimental result is shown in the following figure 1. The precipitable water vapor is used to study changes along the latitude of each station using the average yearly. That is, precipitable water vapor is inversely proportional to latitude, which is similar to what Okulov et al. [1], Gueymard and Garrison [4] and Maghrabi and Clay [12] reported. There were high values for low latitudes and low values at high latitudes, That is, precipitable water vapor is high in the South and low in the North, as shown in figure 2.

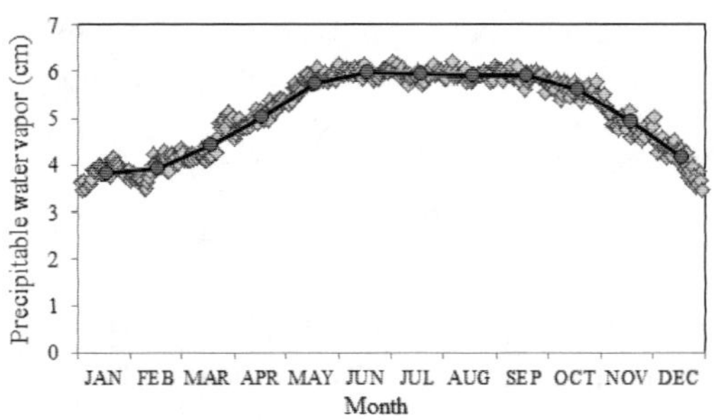


**Figure 1.** Changing patterns according to the time of year of the precipitable water vapor using the average data from 1993 to 2013.

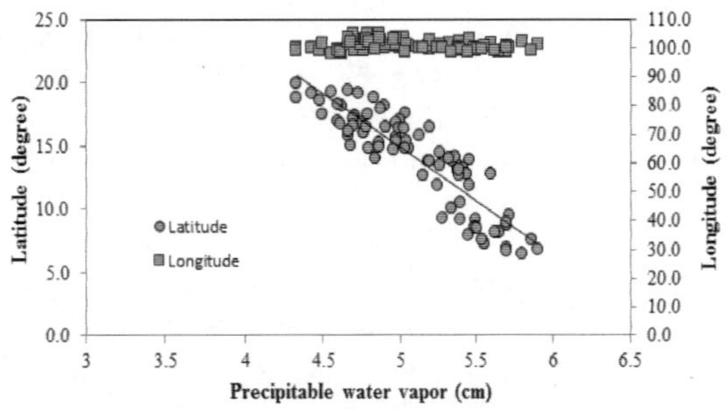


**Figure 2.** Changes along the latitude of the precipitable water vapor using the average annual from surface data of 85 stations during the year 1993-2013.

## 4. Conclusions

From the importance of the amount of water vapor in the atmosphere that affects the weather and climate changes and influencing the intensity of solar radiation. This research has calculated the amount of water vapor from relative humidity and temperature from upper air data of the four stations. The values of the correlation analysis of the temperature, and relative humidity from the surface data, which is measured at the same station. The results of analysis of precipitable water vapor have a relationship with the surface data, which can be written in terms of mathematical models. Test models using independent data, linear regression displayed verify the strong linear relationship between these variables. The study use surface data (temperature and relative humidity) from eighty-five stations across the country, which is applied to the model to calculate the picipitable water vapor. The simulation results showed that the precipitable water vapor changed by the time of the year which is higher in the rainy season from May to October, and less during the dry season from November to April. The precipitable water vapor is high in the South and low in the North.

## 5. Acknowledgements

This work was supported by Rajamangala University of Technology Isan. The author would also like to thank the Department of Meteorology for providing meteorological data.

## References

[1] Okulov O, Ohvril H and Kivi R 2002 Atmospheric precipitable water in Estonia, 1990-2001 *Boreal Env. Res.* **7** 291-300

[2] Kassomenos P A and McGregor G R 2006 The interannual variability and trend of precipitable water over Southern Greece *J. Hydrometerol.* **7** 1–14

[3] Adeyemi B 2009 Emprical modeling of layered integrated water vapor using surface mixing ratio in Nigeria, *J. Appl. Meteorol.* **48** 369-380

[4] Gueymard C A and Garrison J D 1998 Critical evaluation of precipitable water and atmospheric turbidity in Canada using measured hourly solar irradiation *Solar Energy* **62** 291-307

[5] Iziomon M G and Mayer H 2002 Assessment of some global solar radiation parameterizations *Solar Energy* **64** 1631–1643

[6] Nunez M 1993 The development of a satellite-based insulation model for the Tropical Pacific Ocean *J. Climatology* **13** 607-627

[7] Iqbal M 1983 *An Introduction to Solar Radiation* (New York: Academic Press)

[8] Garrison J D 1992 Estimation of atmospheric precipitable water over Australia for application to the division of solar radiation into its direct and diffuse components *Solar energy* **48** 89-96

[9] Leckner B 1978 The spectral distribution of solar radiation at the earth's surface elements of a model *Solar Energy* **20** 143-150

[10] Cohen J L, Salstein D A and Rosen R D 2000 Interannual variability in the meridional transport of water vapor *J. Hydrometeor.* **1** 547–553

[11] Smirnov V V and Moore G W K 2001 Short-term and seasonal variability of the atmospheric water vapor transport through the Mackenzie River basin *J. Hydrometeor.* **2** 441–452

[12] Maghrabi A H and Clay R W 2010 Precipitable water vapour estimation on the basis of sky temperatures measured by a Single-Pixel IR detector and screen temperatures under clear skies *J. Meteorol. Appl.* **17** 279–286

[13] Dai A 2005 Recent climatology, variability, and trends in global surface humidity *J. Climate.* **19** 3589-3606

[14] Bolton D 1980 The computation of equivalent potential temperature *Monthly Weather Review.* **108** 1046-53

# Supersymmetrizing a Quantum Mechanical System

**Patipan Uttayarat**

Department of Physics, Srinakharinwirot University, Wattana, Bangkok 10110, Thailand.

E-mail: patipan@g.swu.ac.th

**Abstract.** Supersymmetry (SUSY) is one of the most active fields of research in high-energy physics and particle physics. However, SUSY is not so easily accessible for students wishing to get started on this field of research due to complexities often associated with quantum field theories. In this article, I'll discuss some aspects of SUSY in the context of quantum mechanics. In particular, I'll demonstrate how to supersymmetrize a familiar quantum system in the case of a square well and a harmonic oscillator. The harmonic oscillator clearly demonstrate the hallmark of SUSY which link a bosonic degree of freedom and a fermionic degree of freedom. I'll briefly discuss the applications of SUSY in particle physics.

## 1. Introduction

Symmetry plays an important role in our understanding of modern physics. For every continuous symmetry there is a conservation law associated with it [1]. To a particle physicist, an object of particular significance is a scattering matrix (S-matrix). Thus it is interesting to identify the maximal symmetry that an S-matrix can possess. In their famous 1967 paper, Coleman and Mandula showed that the maximal symmetry of a non-trivial S-matrix is the product of Poincare symmetry and internal symmetries [2]. However, in 1971 a loophole in the Coleman and Mandula's proof had been identified [3, 4]. This loop-hole allowed for a symmetry generator with spin 1/2 [5] which transforms a boson to a fermion and vice versa. This marked the discovery of supersymmetry (SUSY).

Since its discovery, SUSY has generated enormous research activities. It provides possible solutions to many puzzles in particle physics. First, supersymmetric partners of ordinary particles cure an ultraviolet divergence in the Higgs boson mass, rendering it technically natural [6]. Second, the lightest stable supersymmetric partner could serve as a dark matter [7]. Last, but not least, SUSY allows for the unification of the strong, the weak, and the electromagnetic interactions [8].

Because SUSY is an extension of Poincare symmetry, it's necessarily a symmetry of a quantum field theory (QFT). As a result, SUSY is generically inaccessible for students whom have never taken a course in QFT. However, it is interesting to point out that a QFT in $0 + 1$ dimension is equivalent to a 1-dimensional quantum mechanics. Specifically the operators $x$ and $p$ can be identified with the field operator $\phi(t)$ and its conjugate momentum. Therefore, one could glean some of the features of SUSY in the context of quantum mechanics.

The purpose of this article is to introduce SUSY to a non-particle physicist in a friendly setting of quantum mechanics. Throughout this article, the natural unit system has been employed, ie., $\hbar = c = 1$. For an extensive review on the subject of SUSY in quantum mechanics and its applications, see Ref. [9] and references therein.

## 2. Formulation of SUSY QM

In this section I will formulate SUSY QM in the Hamiltonian framework. Suppose we have a quantum mechanical system $H_1$ that we know its ground state, $\psi_0^{(1)}$. Without loss of generality, we can set the ground state energy to be 0. This gives a relation between the ground state wavefunction and the potential $V_1$ as $V_1(x) = 1/(2m)\,\psi_0''^{(1)}/\psi_0^{(1)}$. Then one can factorize the Hamiltonian as

$$\begin{gathered} H_1 = \frac{p^2}{2m} + V_1(x) = A^\dagger A, \qquad \text{where} \\ A = \frac{ip}{\sqrt{2m}} + W(x) = \frac{1}{\sqrt{2m}}\frac{d}{dx} + W(x), \quad A^\dagger = \frac{-ip}{\sqrt{2m}} + W(x) = -\frac{1}{\sqrt{2m}}\frac{d}{dx} + W(x). \end{gathered} \tag{1}$$

The function $W(x)$ is called the superpotential. It is related to the original potential $V_1$ by $V_1(x) = W^2(x) - W'(x)/\sqrt{2m}$. The superpotential $W(x)$ can also be written in term of the ground state wavefunction $\psi_0^{(1)}$ by noting that $A\psi_0^{(1)} = 0$. Thus $W(x) = -1/\sqrt{2m}\,\psi_0'^{(1)}/\psi_0^{(1)}$.

Now we can define a partner Hamiltonian $H_2 = AA^\dagger$. The spectrum of $H_2$ can be easily obtained from those of $H_1$. Let $\psi_n^{(1)}$ be an eigenstate of $H_1$ with eigenenergy $E_n^{(1)}$. Then

$$AH_1\psi_n^{(1)} = AE_n^{(1)}\psi_n^{(1)} \quad \Longrightarrow \quad H_2A\psi_n^{(1)} = E_n^{(1)}A\psi_n^{(1)}. \tag{2}$$

Thus $A\psi_n^{(1)}$ is an eigenstate of $H_2$ with the same eigenenergy $E_n^{(1)}$. Similarly for an eigenstate $\psi_n^{(2)}$ of $H_2$ with an eigenenergy $E_n^{(2)}$, $A^\dagger\psi_n^{(2)}$ is an eigenstate of $H_1$ with the same eigenenergy. Figure 1 shows the spectra of both $H_1$ and $H_2$.

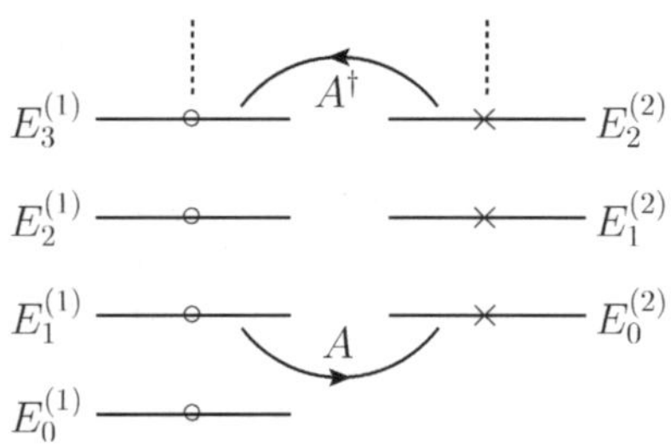


**Figure 1.** The spectra of $H_1$ and $H_2$. The eigenstates of $H_1$ are marked by $\circ$, while those of $H_2$ are marked by $\times$. Note $E_{n+1}^{(1)} = E_n^{(2)}$ and there is no partner state for $\psi_0^{(1)}$ because $A\psi_0^{(1)} = 0$.

### 2.1. SUSY Algebra

The degeneracy in $H_1$ and $H_2$ spectra is the consequence of an underlying SUSY. To see this, we introduce a SUSY Hamiltonian encompassing both $H_1$ and $H_2$, as well as two new operators

$$H = \begin{pmatrix} H_1 & 0 \\ 0 & H_2 \end{pmatrix} = \begin{pmatrix} A^\dagger A & 0 \\ 0 & AA^\dagger \end{pmatrix}, \quad Q = \begin{pmatrix} 0 & 0 \\ A & 0 \end{pmatrix}, \quad Q^\dagger = \begin{pmatrix} 0 & A^\dagger \\ 0 & 0 \end{pmatrix}. \tag{3}$$

It is easy to see that these three operators satisfy

$$[H, Q] = [H, Q^\dagger] = 0, \quad \{Q, Q^\dagger\} = H, \quad \{Q, Q\} = \{Q^\dagger, Q^\dagger\} = 0, \tag{4}$$

which form the SUSY algebra. We can see that $Q$ and $Q^\dagger$ are symmetry generators responsible for the degeneracy in the spectrum of $H_1$ and $H_2$. $Q$ and $Q^\dagger$ are called the supercharge. Moreover, since $Q$ and $Q^\dagger$ are fermionic operators, they change a boson into a fermion and vice versa.

Now consider an eigenstate of $H$ which can be written as $\Psi^n \equiv \begin{pmatrix} \psi_n^{(1)} \\ 0 \end{pmatrix}$ and $\Psi_n \equiv \begin{pmatrix} 0 \\ \psi_n^{(2)} \end{pmatrix}$. The actions of the supercharges are

$$Q\Psi^{n+1} = \begin{pmatrix} 0 \\ A\psi_{n+1}^{(1)} \end{pmatrix} \sim \Psi_n, \quad Q\Psi_n = 0, \quad Q^\dagger \Psi^n = 0, \text{ and } Q^\dagger \Psi_n = \Psi^{n+1}. \tag{5}$$

Since the supercharges are fermionic operators, one can think of $\Psi^{n+1}$ and $\Psi_n$ as having different fermion number. To make this concrete, one can define a fermion number operator. To do this, first note that the ground state of $H$ is $\Psi^0$. This state must have fermion number 0. Thus one can define the fermion number operator as $N_f = \text{diag}(0, 1)$. Hence all the eigenstates $\Psi^n$ have fermion number 0 while the eigenstates $\Psi_n$ have fermion number 1. Notice there is no eigenstate with fermion number greater than 1 because the anticommutation relation of the operator $Q$, see Eq. (4). We will return to this concept in Sec. 3.2.

## 3. Examples

### 3.1. Square Well

Let's consider a familiar infinite square well potential, $V_1(x) = -\pi^2/(2ma^2)$ for $0 \le x \le a$, and $V_1(x) = \infty$ otherwise. The spectrum of $H_1$ can be written as

$$E_n^{(1)} = \frac{n(n+2)\pi^2}{2ma^2}, \quad \psi_n^{(1)}(x) = \sqrt{\frac{2}{a}} \sin\frac{(n+1)\pi x}{a} \quad \text{for } 0 \le x \le a. \tag{6}$$

The superpotential, $W(x)$, and the potential, $V_2(x)$, can then be computed. One gets

$$W(x) = -\frac{1}{\sqrt{2m}}\frac{\pi}{a}\cot\frac{\pi x}{a}, \qquad V_2(x) = \frac{\pi^2}{2ma^2}\left(2\csc^2\frac{\pi x}{a} - 1\right). \tag{7}$$

Finally, one can compute the eigenstate of $H_2$. For example

$$\psi_0^{(2)} \sim A\psi_1^{(1)} \sim \left(\frac{d}{dx}\sin\frac{2\pi x}{a} - \frac{\pi}{a}\cot\frac{\pi x}{a}\sin\frac{2\pi x}{a}\right) \sim \sin^2\frac{\pi x}{a}. \tag{8}$$

The expression for the eigenstate of $H_2$ gets more complicated with increasing value of $n$:

$$\psi_1^{(2)} \sim \sin\frac{\pi x}{a}\sin\frac{2\pi x}{a}, \qquad \psi_2^{(2)} \sim \sin^2\frac{2\pi x}{a} + \sin^2\frac{\pi x}{a}\cos\frac{2\pi x}{a}. \tag{9}$$

One can repeat the same exercise for the finite square well potential $V_1(x) = -V_0$ for $-a \le x \le a$, and $V_1(x) = 0$ otherwise. It's an elementary exercise to determine the ground state energy, $E_0^{(1)}$, and the ground state wavefunction, $\psi_0^{(1)}$. One can then shift the potential $V_1(x)$ so that the ground state energy become zero and apply the steps outlined at the beginning of this section to supersymmetrize the finite square well. Then from

$$\psi_0^{(1)}(x) \sim e^{\sqrt{-2mE_0^{(1)}}|x|} \text{ for } |x| > a, \text{ and } \psi_0^{(1)}(x) \sim \cos\left(\sqrt{2m(V_0 + E_0^{(1)})}\,x\right), \text{ for} |x| \le a, \tag{10}$$

one determines the superpotential, $W(x)$, and the potential, $V_2(x)$, to be

$$W(x) = \begin{cases} -\sqrt{-E_0^{(1)}}, & x < -a, \\ \sqrt{V_0 + E_0^{(1)}}\tan\sqrt{2m(V_0 + E_0^{(1)})}\,x, & -a \le x \le a, \\ \sqrt{-E_0^{(1)}}, & x > a, \end{cases} \tag{11}$$

$$V_2(x) = -E_0^{(1)} \text{ for } |x| > a, \text{ and } V_2(x) = (V_0 + E_0^{(1)})\left[2\tan^2\sqrt{2m(V_0 + E_0^{(1)})}x + 1\right] \text{ for } |x| \le a. \tag{12}$$

### *3.2. Harmonic Oscillator*

For a harmonic oscillator, one can write a Hamiltonian as $H_1 = a_+ a_- \omega$ where $a_+$ is the creation operator and $a_-$ is the annihilation operator. These operators satisfy the algebra $[a_-, a_+] = 1$ and $[a_+, a_+] = [a_-, a_-] = 0$. One can describe the eigenstate of $H_1$ in terms of the bosonic occupation number, $n_b$, as $H_1|n_b\rangle = n_b\omega|n_b\rangle$. Notice that $n_b$ is just an eigenvalue of the bosonic number operator $N_b = a_+ a_-$.

To supersymmetrize the harmonic oscillator, one starts from identifying $A^\dagger = \sqrt{\omega} a_+$ and $A = \sqrt{\omega} a_-$. This allows one to write the Hamiltonian $H_2$ as $H_2 = AA^\dagger = H_1 + \omega$. Then one can define the SUSY Hamiltonian and the supercharges as in Eq. (3). The eigenstate of $H$ are $\Psi^n$ and $\Psi_n$ as defined in Sec. 2.1. One can label the eigenstate of the SUSY harmonic oscillator in terms of its boson number and fermion number, $|n_b, n_f\rangle$. For example, $|0,0\rangle = \Psi^0$, $|1,0\rangle = \Psi^1$, $|0,1\rangle = \Psi_1$, etc. Hence one can see that $Q|n_b, n_f\rangle \sim |n_b - 1, n_f + 1\rangle$ and $Q^\dagger|n_b, n_f\rangle \sim |n_b + 1, n_f - 1\rangle$. This makes explicit the fact that a SUSY transformation turns a bosonic degree of freedom into a fermionic one and vice versa.

## 4. Discussion and Final Remarks

SUSY predicts that there should be a superpartner of the electron, selectron. The selectron has spin 0, electric charge -$e$ and mass $m_e$. However, such a particle hasn't been observed in nature. Thus if SUSY is to be a symmetry of nature, it must be spontaneously broken. In fact, SUSY QM was originally introduced as a toy model for studying SUSY breaking [10, 11] to avoid unnecessary complications of QFT. In addition to serving as a toy model for studying aspect of SUSY in QFT, SUSY QM generates a large areas of research activity of its own. Many of them are reviewed in Ref. [9].

In the context of QFT, SUSY has found many application in particle physics. In the minimal supersymmetric Standard Model (MSSM) of particle physics, every particle comes with a superpartner. In MSSM the lightest superpartner (LSP) is stable. If in addition the LSP is neutral, it can serve as a dark matter candidate. Moreover, interactions of the Higgs boson with superpartners cancel the quantum corrections to the Higgs boson mass from ordinary particles. In the absence of quantum corrections, the Higgs mass can naturally be at 125 GeV as measured by the ATLAST and CMS experiments [12, 13]. Lastly, the quantum corrections lead to renormalization [14, 15]. Renormalization implies the strength of the strong, the weak and the electromagnetic interactions change with energy scale. As it turns out, in the MSSM, renormalization unifies the three interaction at the high energy scale, dubbed the grand unification scale.

## Acknowledgments

The work of PU has been supported in part by the Thailand Research Fund under contract no. TRG5880061, and the Faculty of Science, Srinakharinwirot University under grant no. 655/2559.

## References

[1] Noether E 1918 *Gott. Nachr.* **1918** 235–257 [Transp. Theory Statist. Phys.1,186(1971)]
[2] Coleman S R and Mandula J 1967 *Phys. Rev.* **159** 1251–1256
[3] Golfand Yu A and Likhtman E P 1971 *JETP Lett.* **13** 323–326 [Pisma Zh. Eksp. Teor. Fiz.13,452(1971)]
[4] Ramond P 1971 *Phys. Rev.* **D3** 2415–2418
[5] Haag R, Lopuszanski J T and Sohnius M 1975 *Nucl. Phys.* **B88** 257
[6] 't Hooft G 1980 *NATO Sci. Ser. B* **59** 135–157
[7] Jungman G, Kamionkowski M and Griest K 1996 *Phys. Rept.* **267** 195–373
[8] Dimopoulos S, Raby S and Wilczek F 1981 *Phys. Rev.* **D24** 1681–1683
[9] Cooper F, Khare A and Sukhatme U 1995 *Phys. Rept.* **251** 267–385
[10] Witten E 1981 *Nucl. Phys.* **B188** 513

[11] Cooper F and Freedman B 1983 *Annals Phys.* **146** 262
[12] Aad G *et al.* (ATLAS) 2012 *Phys. Lett.* **B716** 1–29
[13] Chatrchyan S *et al.* (CMS) 2012 *Phys. Lett.* **B716** 30–61
[14] Callan Jr C G 1970 *Phys. Rev.* **D2** 1541–1547
[15] Wilson K G and Kogut J B 1974 *Phys. Rept.* **12** 75–200

IOP Conf. Series: Journal of Physics: Conf. Series **901** (2017) 012053 doi:10.1088/1742-6596/901/1/012053

# Forbush decreases detected at the Princess Sirindhorn Neutron monitor in the 24$^{th}$ Solar Cycle

**M Deeprom[1,*] and T Nutaro[1]**
[1]Department of Physics, Faculty of Science, Ubon Ratchathani University, Ubon Ratchathani, Thailand.

[*] E-mail: maliwan.de.57@ubu.ac.th

**Abstract**. Neutron monitors (NMs) are ground-based detectors designed to measure the cosmic ray (CR) intensity by measuring secondary particles. A Forbush decrease (FD) is a sudden decrease in the CR-intensity count rate and the amplitude of the decreases changes with the different cutoff rigidity of each NM station. The Princess Sirindhorn NM (PSNM) was established in 2007, an NM64-type and 3 bare tubes located at the world's highest geomagnetic vertical cutoff rigidity 16.8 GV was installed at Doi Inthanon in Thailand, 2565 m above sea level. In this work, we study the FDs in CR intensity detected at PSNM that occurred during the 5 years of 2010-2015 in the 24$^{th}$ solar cycle. We analyse the FDs at the Oulu NM in Finland in order to identify and compare them. We identify 46 FDs at both PSNM and Oulu NMs then we used the student's t-test to assess difference in the mean between simultaneous or non-simultaneous of FDs event. We found that the 46 FDs, 22 are simultaneous and 24 non-simultaneous. Furthermore, we also found that student's t-test analysis reveals that FDs at PSNM have smaller intensity variation of main phase duration of main phase than at Oulu NM. The results of the present statistical analysis that support the PSNM can provide CR data of reliability comparable to that of the Oulu NM. This work also provides unique data of FDs and solar modulation.

## 1.Introduction

Forbush decrease (FD) [1] represents a sharp decrease in the intensity of galactic cosmic ray (GCR), followed by disturbance in the strong solar events and usually associated with passage of interplanetary coronal mass ejection (ICME) [2, 3]. FDs can be observed on Earth by using ground-based particle detectors. Neutron monitors (NMs) are ground-based detectors designed to estimate the primary cosmic ray (CR) intensity by measuring secondary particles. These CR decreases are recorded on Earth by NM stations of the worldwide network. The amplitude of the decreases changes with the different cutoff rigidity of each NM stations. The Princess Sirindhorn NM (PSNM) was installed at DoiInthanon, Thailand (18.59N, 98.49E, 2565 m altitude) where the effective cutoff rigidity is 16.8 GV [4]. PSNM provides a basis for comparison with count rates of other NMs worldwide which are similarly calibrated. In this work, we examine simultaneous and non-simultaneous FDs using the CR data observed by PSNM in Thailand compared with Oulu NM (OULU) in Finland that occurred during the five years of 2010-2015 in the 24th Solar Cycle. We analyse the statistical properties of CR data observed by both NM stations for simultaneous and non-simultaneous FDs.

## 2.Data and Selection of FD

FD consists of a main phase and recovery phase. In the main phase, there is a rapid decrease in the intensity of CR. The onset of the main phase is the time that the maximum intensity is recorded before the CR- intensity begins to decrease. The end of the main phase is the time of minimum CR- intensity. The onset time is determined within 6 h from onset of the decrease, and the end time is determined within 12 h from the termination of the decrease [5]. We used the hourly pressure-coerced values of the CR intensity observation data during from 2010-2015 at the PSNM and OULU. The variation in CR-intensity is normalized to the zero level of the pre-event, the CR-intensity to the monthly average for each NM station. In this study, we selected the well-defined FDs with an amplitude >3% at the OULU. Then we find the FDs at the PSNM that correspond to those at the OULU. The simultaneity of FDs are generally associated with ICMEs, thus we consider FDs due to the passage of ICMEs such as magnetic cloud and ejecta, the ICME list by Ian Richardson and Hilary Cane [6]. In this work, we used the criteria to select of simultaneous and non-simultaneous of FDs suggested by Lee SS et al., (2015). When the decrease in the CR intensity of main phase on the FD profile of NM overlapped in Universal Time (UT), the FDs were classified as simultaneous FDs and if the FDs that non-overlap in UT were classified as non-simultaneous FDs [7].

**Table 1.** Statistical properties of simultaneous and non-simultaneous FD events.

| **NM** | Statistical properties | SFD [a] | NSFD [b] | Total | Confidence level (%) of $I_S > I_{NS}$ [c] |
|---|---|---|---|---|---|
| | Number of FDs | 22 | 24 | 46 | |
| PSNM | Mean of CR intensity - | 2.3±0.8 | 1.7±0.4 | 2.0±0.7 | 99.6% |
| OULU | during main phase (%) | 5.0±1.9 | 4.1±0.9 | 4.5±1.5 | 96.4% |
| PSNM | Duration time | 19.7±9.7 | 14.9±5.7 | 17.2±8.2 | |
| OULU | of main phase (h) | 18.7±8.1 | 12.9±3.4 | 15.7±6.7 | |

[a] simultaneous Forbush decrease,
[b] Non-Simultaneous Forbush decrease.
[c] CR-intensity variation of main phase

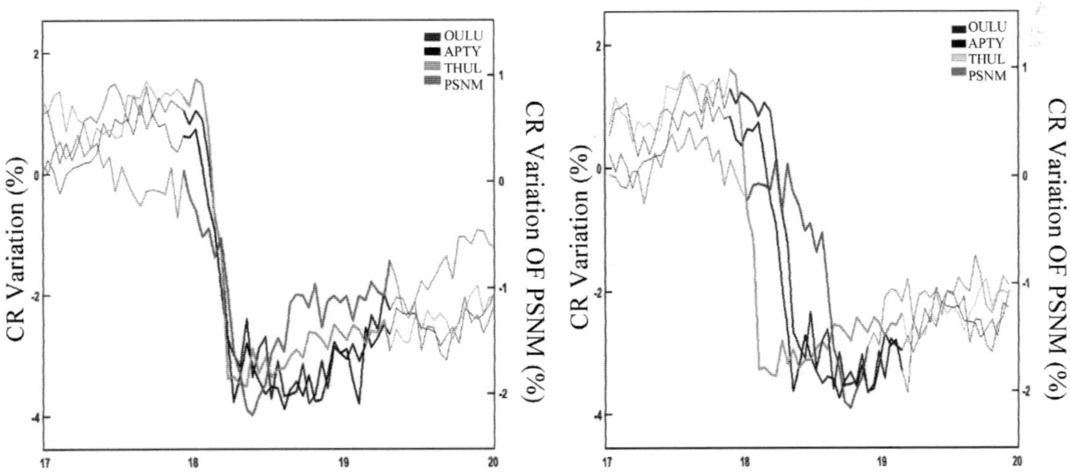


**Figure 1.** CR- intensity variation profiles of a simultaneous FD event on 18 February 2011 observed at four NM station in (a) UT and (b) LT.

## 3.Results

Table 1 shows the statistical properties of the FDs, including the number of FDs, the mean of the CR - intensity and the result Student's t-test. We used the Student's t-test to assess differences in the mean between the types of FDs. Of the 46 FDs, 24 FDs were classified as simultaneous FDs, and 22 FDs as non-simultaneous. The intensity variation of main phase at the PSNM is smaller than the variation at the OULU and the mean of CR-intensity variation of the simultaneous events are higher than the variation from non-simultaneous events at both NM stations by Student's t-test. The results from the student t-test show that, these difference between two types of events is statistically significant with 95% confidence. Figure 1 shows a typical example for the CR-intensity profile of a simultaneous FD on 18 February 2011 observed at four NMs with different vertical cufoff rigidities. All NM stations have the main phase of FD events overlapped on the CR intensity profile based in UT (Figure1a). According to that the FD events of 18 February 2011 has a simultaneous FD was recorded at all NM around the world [8]. Figure 1b show the CR-intensity variation of both NM stations based in LT. Whilst in LT the main phases occur at different times at difference NM stations, the main phases of FD did not overlap, due to the difference. Figure 2 shows for the CR-intensity profile of a non-simultaneous FD that occurred on event on 18 April 2014. These phenomenon is caused by the differences in time imposed by longitude of NM stations.

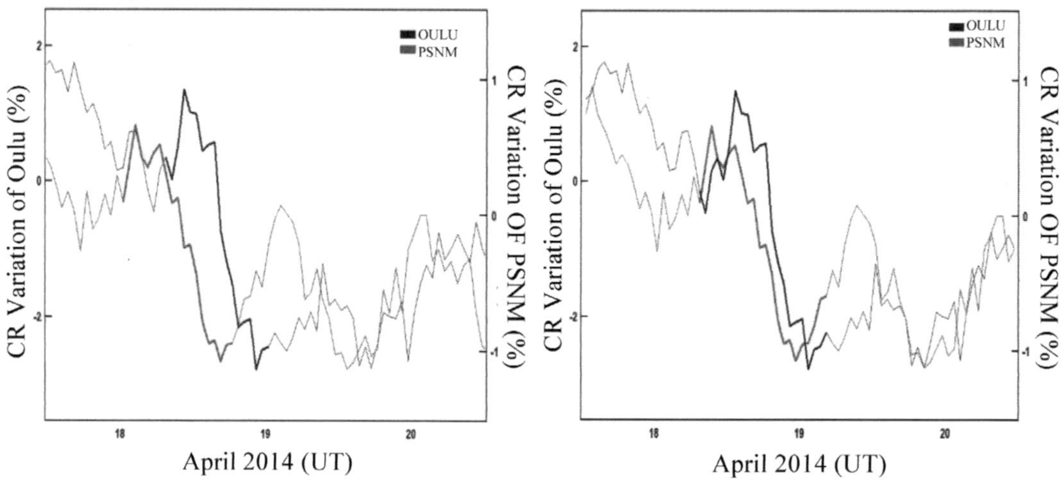


**Figure 2.** CR- intensity variation profiles of a non-simultaneous FD event on 18 April 2014 observed at the PSNM and OULU NM stations in (a) UT and (b) LT.

Figure 3 shows the distributions and the diurnal variation of onset time of the main phase in LT. Figure 3a shows of the CR-intensity, for FDs recorded at PSNM in the polar coordinates. The unfilled circle and filled mark the onset times and mean CR-intensity observed from both stations. From the plot, most non-simultaneous events have onset time on dayside and most simultaneous events have higher CR-intensity decrease. The maximum intensity of diurnal variation manifests at 12 LT at PSNM and 13 LT at OULU as shown in figure 3b, that the distribution for onset time in LT at PSNM of non-simultaneous FDs peaks near 12 LT.

## 4.Conclusion

In this paper, we reported the results of our study of FDs observed at PSNM, the world's highest magnetic cutoff rigidity 16.8 GV. In order to consider the reliability of CR data recorded by OULU (Finland), we selected FDs that occurred during 5 years between 2010-2015 in the $24^{th}$ Solar Cycle. We repeated analysis to test the FDs at the PSNM in order to identify and compare with the OULU. We also classified into simultaneous and non-simultaneous FD events using overlapping main phase on the cosmic ray profile in UT. The study results are summarized as follows.

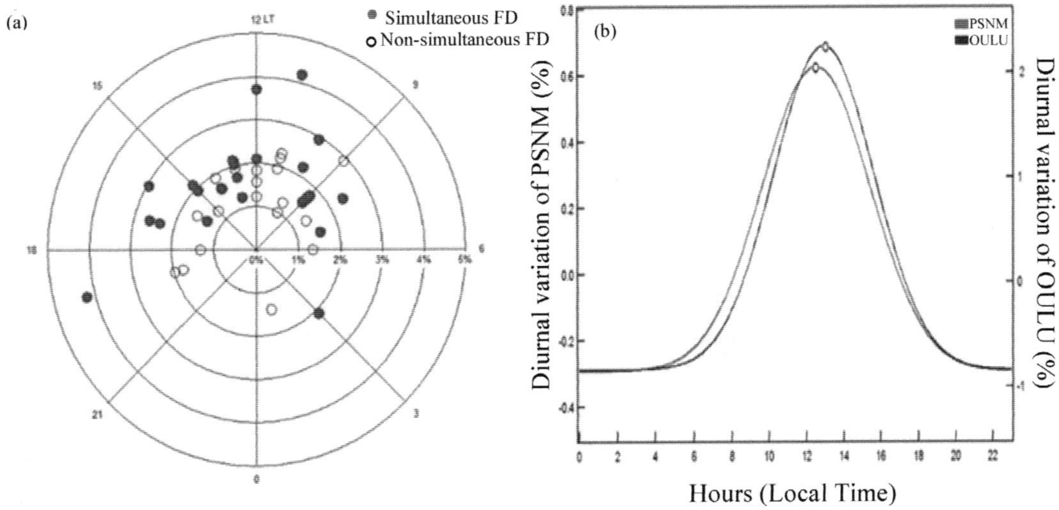


**Figure 3.** Distributions of the onset time (LT) of the FD main phases at PSNM. Figure a, the radial shows the magnitude of the CR-intensity variation, and the angular coordinate component show the LT at PSNM (lift). Figure b, the diurnal variation at both NM stations (red at PSNM and blue of OULU) during the analysis period (right).

(1) We identify during the period of analysis, 46 FD events occurred, 22 are simultaneous FDs and 24 FD events are non-simultaneous FDs. The CR intensity variation is larger in of the simultaneous FD events, than the non-simultaneous FD events at the PSNM and OULT NM stations with very high confidence levels in the student t-test.

(2) Of the 46 FD events during the analysis period at the PSNM and the OULT. Student t-test results indicate FD events at the PSNM have smaller CR intensity variation of main phase, while the duration of the main phase longer than do those at the OULU. The main phase onset time of non-simultaneous FD event distributed mostly within 12-hour intervals, in dayside onset time, in line with the typical properties of non-simultaneous FDs guidance by Kang J Set al. and Oh S Yet al. [9, 10].

**Acknowledgments**
We would like to thank the Oulu station for data providing by the NMDB database (www.nmdb.eu). Many thanks are also to the Princess Sirindhorn Neutron Monitor (PSNM) for the whole data since the station was installed.

**References**
[1] Forbush S E 1938 *J. Geophys. Research* **43** 203–218
[2] Cane H V 2000 *Space Sci. Rev.* **93** 55–77
[3] Lockwood J A 1971 *Space Sci. Rev.* **12** 658–715
[4] Ruffolo D 2011, *Proc. of 32nd Int. Cosmic Ray Conf. (Beijing)* vol 10 pp 265–267
[5] Lee S S, Oh S Y, and Yi Y 2013 *J. Geophys Research* **118** 608–614
[6] http://www.srl.caltech.edu/ACE/ASC/DATA/level3/icmetable2.htm#(e)
[7] Lee S S et al., 2015 *J. Astoron Space Sci.* **32** 33-38
[8] Papaioannou A et al., 2013 *J. Phys.: Conf. ser.* **409** doi:10.1088/1742-6596/409/1/012202
[9] Kang J S, Oh S Y, Kim Y K, 2015, *Adv. Space Res.* **57** 912–918
[10] Oh S Y, Yi Y, and Kim Y H, 2008 *J. Geophys. Research.* **113**, A0110

IOP Conf. Series: Journal of Physics: Conf. Series **901** (2017) 012057 doi:10.1088/1742-6596/901/1/012057

# Characteristics of Linear Variable Differential Transformer(LVDT) Probe for Gauge Blocks Calibration

**S Peingbangyang[1] and J Wongsaroj[1]**

[1] Length Laboratory, Department of Dimensional Metrology, National Institute of Metrology (Thailand), 3/4-5 Moo 3, Klong 5, Klong Luang, Pathumthani 12120, Thailand

Corresponding author : samana@nimt.or.th

**Abstract.** The LVDT probe is a very important component used in gauge blocks calibration via a mechanical comparative method. The probe is used to determine the central length difference between a reference gauge block and gauge block under-tested (UTC). Typically, an UTC and a reference gauge block have the same nominal length. However, some gauge blocks UTC are specially made for specific purpose where the reference gauge blocks with the same nominal length are not commercially available. Various reference gauge blocks are wrung in order to provide nominal length the same as that of the UTC. Wringing process is the troublesome step and causes larger measurement uncertainty. The lower accuracy is higher number of gauge block used to create reference gauge block, in order to improve accuracy of measurement, the LVDT probe was used at the longer range where the reference gauge block and the UTC don't need to be the same nominal length. In this paper, characteristics of LVDT probe was investigated as it is related to the accuracy of the measurement result. Errors of LVDT probe came from non-linearity, calibration factor, retrace error, repeatability and maximum difference in length. A pair of gauge block, calculated by the Twyman-Green interferometer, length different range 5 μm to 230 μm was used in the study. Non-linearity of LVDT is evaluated by a simple linear regression model. The non-linearity of LVDT probe, calibration factor, retrace error, repeatability and maximum difference in length are 25 nm, 1.0003, 3 nm, 6 nm and 80 nm, respectively. Therefore, by using this technique, central length difference between the 2 gauge blocks up to 80 nm can be calibrated with the uncertainty due to non-linearity of 15 nm The experiment shows the large error of retrace closes to 0.02 μm at 0.09 μm. This can be determined the maximum difference in length to calibrate gauge blocks at difference nominal lengths. The measurement uncertainty of non-linearity is evaluated and it is close to 15 nm.

## 1. Introduction

Gauge block is used to calibrate dimensional measuring instrument such as micrometer, vernier caliper, coordinate measuring machine (CMM) and universal length machine (ULM) because it is easy to use and with high accuracy. Gauge block calibration generally follows the standard ISO3650:1998 [1] which is using mechanical comparison method. In order to determine length of an unknown gauge block, the difference of its central length from that of the reference gauge block shall be measured. After applying algerbraically to the actual length of the reference gauge block, true length of an unknown gauge block can be determined. For probing, the measuring faces of each gauge are approached from opposite

IOP Conf. Series: Journal of Physics: Conf. Series **901** (2017) 012057 doi:10.1088/1742-6596/901/1/012057

directions and length difference is measured by a high resolution indicator in the gauge block comparator system.

## 2. Measurement principle

### *2.1. Central length measurement*

The calibration of gauge block is carried out by comparison using comparator and a calibrated gauge block of the same nominal length and the same material as reference standard as show in figure 1.

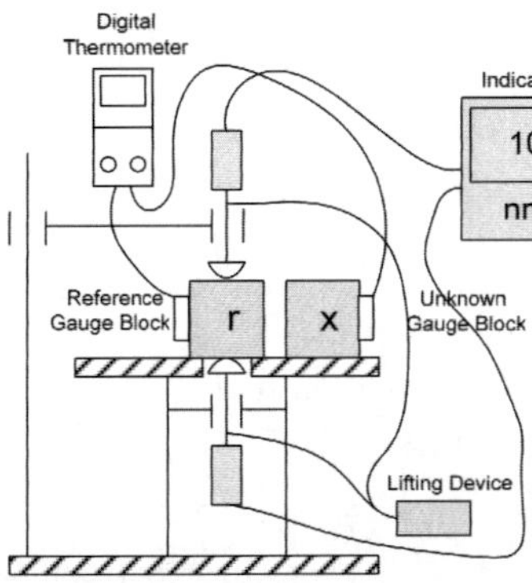


**Figure 1**. Schematic diagram of gauge block calibration

Error sources of gauge block calibration by mechanical comparison are repeatability, reference gauge block, temperature correction and non-linearity of LVDT probes. For a dual probe gauge block comparator, the central length of unknown gauge block ($l_x$) at the reference temperature is obtained from the relationship in equation (1) [2].

$$l_x = l_s + \delta l_D + \delta l + \delta l_C - L\left(\bar{\alpha} \times \delta t + \delta\alpha \times \Delta\bar{t}\right) - \delta l_V \quad (1)$$

Where; $l_s$ is length of the reference gauge block at the reference temperature $t_0$=20 °C according to its calibration certificate, $\delta l_D$ is change of the length of the reference gauge block since its last calibration due to drift, $\delta l$ is observed difference in length between the unknown and the reference gauge block, $\delta l_C$ is correction for non-linearity and offset of the comparator, $L$ is nominal length of the gauge blocks considered, $\bar{\alpha} = (\alpha_x + \alpha_s)/2$ is average of the thermal expansion coefficients of the unknown and reference gauge blocks, $\delta t = t_x - t_s$ is temperature difference between the unknown and reference gauge blocks, $\delta\alpha = \alpha_x - \alpha_s$ is difference in the thermal expansion coefficients between the unknown and the reference gauge blocks, $\Delta\bar{t} = (t_x + t_s)/2 - t_0$ is deviation of the average temperature of the unknown and the reference gauge blocks from the reference temperature and $\delta l_V$ is correction for non-central contacting of the measuring faces of the unknown gauge block.

*2.2. Non-linearity measurement and calibration factor of LVDT*
Correction for non-linearity of LVDT probes in equation (1) is main error sources when unknown and reference gauge blocks are difference nominal length. Length different of a pair of gauge block calculated by the Twyman-Green interferometer ($l_C$) was compared with length different measuring by the comparator ($l_M$) for non-linearity measurement. Non-linearity of LVDT probes is evaluated by a simple linear regression model [3].

$$l_{CM} = k_0 + k_1 l_C \qquad (2)$$

Where; $l_{CM} = l_C - l_M$, $k_0 = \bar{l}_{CM} - k_1 \bar{l}_C$ and $k_1 = \frac{\sum_{i-1}^{n} \left(l_{C,i} - \bar{l}_C\right)\left(l_{CM,i} - \bar{l}_{CM}\right)}{\left(l_{C,i} - \bar{l}_C\right)^2}$

Non-linearity correction of LVDT probes was calculated by $\delta l_C = \left(l_{CM,\max} - l_{CM,\min}\right)/2$ , uncertainty due to non-linearity was calculated by $u(\delta l_C) = \delta l_C / \sqrt{3}$ [4] and calibration factor ($C$) was determined by the ratio between length different calculated by the Twyman-Green interferometer ($l_C$) and length different measuring by the comparator ($l_M$).

## 3. Preliminary experiment and discussion
The experiment was carried out on dual probes gauge block comparator (GBC) manufacturer by Mahr, Model 826 PC. Schematic diagram of the GBC is illustrated in figure 2. GBC consist of a high resolution display unit, upper probe and lower probe made of tungsten carbide with ball the diameter of 3 mm and measuring force of dual 0.75 N and temperature compensation unit manufacturer by Albhorn, Model 2390-8.

**Figure 2**. Experimental setup

A pair of gauge block made of steel manufacturer by Mitutoyo, calculated by the Twyman-Green interferometer, length different rage 5 μm to 230 μm was used in the study. Table 1 show the measurement results of LVDT probes.

**Table 1.** Measurement results of LVDT probes

| $l_C$ (μm) | $l_M$ (μm) | $l_{CM}$ (μm) | $C$ |
|---|---|---|---|
| -230.036 / 230.036 | -229.847 / 229.870 | -0.189 / 0.166 | 1.0008 / 1.0007 |
| -200.022 / 200.022 | -199.873 / 199.883 | -0.149 / 0.139 | 1.0007 / 1.0007 |

| $l_C$ (μm) | $l_M$ (μm) | $l_{CM}$ (μm) | $C$ |
|---|---|---|---|
| -180.053 / 180.053 | -179.920 / 179.923 | -0.133 / 0.130 | 1.0007 / 1.0007 |
| -100.023 / 100.023 | -99.963 / 99.967 | -0.060 / 0.056 | 1.0006 / 1.0006 |
| -90.063 / 90.063 | -89.983 / 90.003 | -0.080 / 0.060 | 1.0009 / 1.0007 |
| -80.010 / 80.010 | -79.983 / 79.987 | -0.027 / 0.023 | 1.0003 / 1.0003 |
| -70.007 / 70.007 | -69.983 / 69.990 | -0.024 / 0.017 | 1.0003 / 1.0002 |
| -60.001 / 60.001 | -59.977 / 59.983 | -0.024 / 0.018 | 1.0004 / 1.0003 |
| -50.024 / 50.024 | -50.000 / 50.003 | -0.024 / 0.021 | 1.0005 / 1.0004 |
| -39.990 / 39.990 | -39.980 / 39.990 | -0.010 / 0.000 | 1.0003 / 1.0000 |
| -30.049 / 30.049 | -30.030 / 30.037 | -0.019 / 0.012 | 1.0006 / 1.0004 |
| -20.038 / 20.038 | -20.037 / 20.043 | -0.001 / -0.005 | 1.0001 / 0.9997 |
| -9.909 / 9.909 | -9.907 / 9.910 | -0.002 / -0.001 | 1.0002 / 0.9999 |
| -4.784 / 4.784 | -4.783 / 4.783 | -0.001 / 0.001 | 1.0001 / 1.0001 |

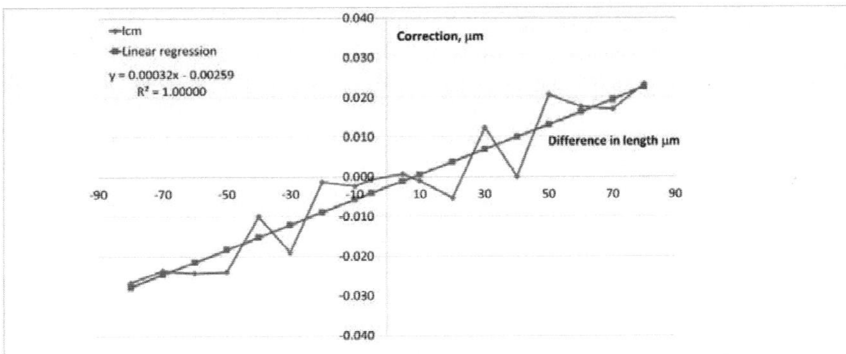


**Figure 3**. Non-linearity of LVDT probes

The non-linearity of LVDT probe in figure 3, calibration factor, retrace error, repeatability and maximum difference in length are 25 nm, 1.0003, 3 nm, 6 nm and 80 nm respectively. Therefore, by using this technique, central length difference between two gauge blocks up to 80 nm can be calibrated with the uncertainty due to non-linearity of 15 nm. The experiment shows large error of retrace closes to 0.02 μm at 0.09 μm. This can be determined the maximum difference in length to calibrate gauge block at difference nominal length.

## 4. References

[1] International Organization for Standardization 1998 *ISO 3650 Geometrical Product Specifications(GPS) – Length Standards – Gauge Blocks* Second edition 11-12
[2] Europian Accreditation 2013 *Evaluation of the Uncertainty of Measurement in Calibration* September 2013 rev 02 32-35
[3] E. Kreysig 2011 *Advance Engineering Mathematics* tenth edition New Jersey 1103-11035
[4] A. Godina, B. Acko and M. Druzovec 2007 *New approach to uncertainty evaluation in the calibration of gauge block comparators* Measurement 40 (2007) 607-614

IOP Conf. Series: Journal of Physics: Conf. Series **901** (2017) 012058 doi:10.1088/1742-6596/901/1/012058

# A New Proficiency Testing (PT) Evaluation for Conformity Assessment of Roughness Measuring Instruments

**A Teppitaksak* and J Buajarern**

Department of Dimensional Metrology, National Institute of Metrology Thailand, 3/4-5 Moo 3, Klong 5, Klongluang, Pathumthani, 12120 Thailand

*Corresponding author: achayat@nimt.or.th

**Abstract**. Precise surface roughness measurement is crucial as it has a direct effect on surface quality. In this work, Proficiency Testing (PT), which is an important process for conformity assessment, was conducted based on a portable roughness tester as a standard measuring instrument. There were seven calibration laboratories from Thailand that participated in the PT with a technical protocol that was designed by NIMT. The roughness tester was provided by NIMT. Two roughness standards with different roughness values (Ra ≤1 and Ra >1) were prepared by each calibration laboratory. The PT measurands were common roughness parameters such as Ra and Rsm. As the roughness measuring instrument is based on LVDTs, we propose here a new proficiency testing evaluation which involves the calibration of the measuring instrument in the PT evaluation procedure. The PT results, according to ISO/IEC 17043, shows that six out of the seven participating laboratories indicated satisfactory performance ($|En| \leq 1$) where the measurement uncertainty was based on a 95% confidence level. The PT results and evaluation procedure for conformity assessment of the roughness measuring instrument are presented. Finally, measurement techniques to improve measurement quality are recommended.

## 1. Introduction

Surface finishing is an important process for manufacturing industries. Accurate surface roughness measurement is crucial as it has a direct impact on quality of machine parts. In Thailand, the National Institute of Metrology of Thailand (NIMT) provides traceability of reference roughness standards and roughness measuring instruments. NIMT is also a coordinator to organize proficiency testing (PT), which is an important process for conformity assessment [1]. Previously, PT of surface roughness measurement has been focusing on roughness specimen [2]. Roughness specimens are typically used as a reference standard for calibrating a portable roughness measuring instrument (roughness tester). A roughness tester is widely used in manufacturing industries due to its advantages such as compactness and lightness. However, uncertainty of roughness tester calibration is not only arises from reference roughness standards. There are other sources of uncertainties such as environmental errors, human errors, setting of measuring condition such as cut-off wavelength, evaluation length [3-4]. Therefore, PT on roughness tester calibration is also important. In this work, we organized a PT based on a portable roughness tester. PT evaluation for conformity assessment of the roughness tester including method, results, analysis and discussion will be presented. A new evaluation method that utilizes a linearity of a stylus probe of the roughness tester in the calibration process will be discussed in detail.

## 2. Methodology

Surface laboratory of NIMT organized the PT using roughness tester (Surtronic S128 by Taylor Hobson) as an artifact. There were seven laboratories participated in the program. All participants required to prepare at least two high accuracy roughness standards with nominal value of Ra ≤1 and Ra >1. Traceability evidence of roughness standards i.e. calibration certificates must be provided. Cutoff wavelength and bandwidth setting were recommended to follow ISO4288:1996 using Gaussian filter. Participants were advised to measure roughness parameters according to laboratory procedures, which should include Ra and Rsm values. Noted that all participants were informed not to use "calibration" function to calibrate the roughness tester.

## 3. Results

PT results from NIMT and all participants are shown in Table 1. Participants laboratories are named by code A-G. Reference values shown in the table are roughness standard values used in the PT by each laboratories, which obtained from the certificates. Measurement uncertainties, shown in blacked after the results, are based on 95% level of confidence (k=2).

| | Ra | | Rsm | |
|---|---|---|---|---|
| **Code** | **Reference value** | **Measured value** | **Reference value** | **Measured value** |
| NIMT-1 | 0.834(0.04)<br>3.125(0.05) | 0.771(0.142)<br>3.281(0.478) | 80(0.58)<br>100(0.58) | 89.640(2.674)<br>108.200(3.135) |
| A | 0.398 (0.04)<br>2.892(0.04) | 0.390(0.083)<br>3.010(0.174) | 15(0.58)<br>94(0.58) | 16.000(2.044)<br>101.000(12.140) |
| B | 0.41(0.06)<br>2.911(0.08) | 0.388(0.058)<br>2.980(0.198) | 15(0.58)<br>94(0.58) | 16.204(1.038)<br>102(6.554) |
| C | 2.991(0.037)<br>0.476(0.037) | 2.933(0.103)<br>0.483(0.092) | N/A<br>N/A | N/A<br>N/A |
| D | 2.942(0.02)<br>0.459(0.03) | 2.250(0.430)<br>0.470(0.420) | 100(0.58)<br>56.5(8.79) | 109.000(0.990)<br>35.000(14.000) |
| E | 0.403(0.01)<br>3.163(0.04) | 0.350(0.080)<br>3.270(0.185) | N/A<br>N/A | N/A<br>N/A |
| F | 0.399(0.01)<br>2.967(0.04) | 0.376(0.083)<br>3.014(0.534) | N/A<br>N/A | N/A<br>N/A |
| G | 5.99(0.142)<br>0.83(0.023) | 6.070(0.154)<br>0.806(0.154) | N/A<br>N/A | N/A<br>N/A |
| NIMT-2 | 0.0834(0.04)<br>3.125(0.05) | 0.753(0.143)<br>3.279(0.533) | 80(0.58)<br>100(0.58) | 89.040(2.762)<br>106.760(3.279) |

Table 1: Summary of laboratory measurement results in μm.

## 4. Analysis and Discussion

Conventional statistical evaluation methods such as z-scores and En ratio rely on a reference laboratory's value to evaluate participating laboratories. However, in this PT, participants prepared their own roughness standards. It is, therefore, impractical to evaluate participating laboratories performance based on a direct comparison with NIMT as a reference laboratory result.

In this PT program, a roughness value for each roughness standard was obtained from its calibration certificate and used as a reference value. As the portable roughness tester is based on LVDTs, calibration factor (or correction factor) was calculated to correct the measured results. The calibration factor (Cf) is determined as,

$$C_f = \frac{\upsilon_{ref}}{\upsilon_{mea}} \qquad \text{Eq.1}$$

where $\upsilon_{ref}$ is the reference value obtained from calibration certificate and $\upsilon_{mea}$ is the value measured using roughness tester. By plotting $\upsilon_{ref}$ against $\upsilon_{mea}$, a correction factor was obtained from a slope of the curve. Assuming a linear relationship between υmea and υmea at all ranges, a correction factor based on Rsm and Ra for each laboratory was calculated based on linear regression using least square method, as shown in table 2. Note that the correction factor of NIMT and participant lab codes A,B and E-G are lower than 1, while the correction factor of lab C and D are higher than 1. This indicates opposite correcting direction of the measured values. In other words, the corrected value, or the measured value multiplied by it's correction value, of lab A-B and E-G will increases while lab C and D will decreases compared on measured value.

An inaccurate correction factor could result from human error of an operator or the quality of the reference standard. Cut-off wavelength setting, sampling length setting and improper alignment could also lead to error in the correction factor. These things were found to be at least partly responsible for the differences in the correction factors obtained from labs C and D compared to the other labs. To improve measurement quality, the cutoff wavelength and bandwidth setting must follow the standard described in ISO4288:1996.

| Code | Correction factor | |
|---|---|---|
| | Rsm | Ra |
| NIMT-1 | 0.9113 | 0.9592 |
| A | 0.9309 | 0.9618 |
| B | 0.9182 | 0.9782 |
| C | N/A | 1.0189 |
| D | 0.9761 | 1.2937 |
| E | N/A | 0.9694 |
| F | N/A | 0.9856 |
| G | N/A | 0.9876 |
| NIMT-2 | 0.9210 | 0.9607 |

Table 2. Summary of correction factor of Rsm and Ra.

The performance of the participants was evaluated based on normalized error (En), according to [1],

$E_n = \frac{\upsilon_{result} - \upsilon_{ref}}{\sqrt{U_{lab}^2 + U_{ref}^2}}$ ,where $\upsilon_{result}$ is the corrected measurement result ($\upsilon_{result} = \upsilon_{mea} \times C_f$), $U_{lab}$ is the participant's expanded uncertainty at a 95% level of confidence and $U_{ref}$ is the expanded uncertainty from a calibration certificate at a 95% level of confidence. According to ISO/IEC 17043[1], |En| ≤ 1 indicates "satisfactory" degree of compatibility between the participant's result and the reference value, and is considered to be internationally acceptable. |En| outside this range indicates "unsatisfactory" performance which requires immediate investigation and corrective action. Figure 1 shows a Summary of corrected value of Ra and Rsm from all participants. As shown in Fig.1, |En| value shows generally good agreement between participants and the reference value. For labs A-C and E-G, the En values of Ra and Rsm are ≤ 1. Only two |En| values of Ra and Rsm were greater than 1, these were both from lab D.

In this PT, "unsatisfactory" |En| values could arise from various factors such as (i) inaccurate correction factor as described above (ii) misalignment of the stylus probe to the roughness standard

(iii) any other errors not included in the calculation of uncertainty. It was noted that most of the participating labs reported lower uncertainty than NIMT. This indicated that some sources of uncertainty were neglected. Therefore, further training of uncertainty calculation is recommended.

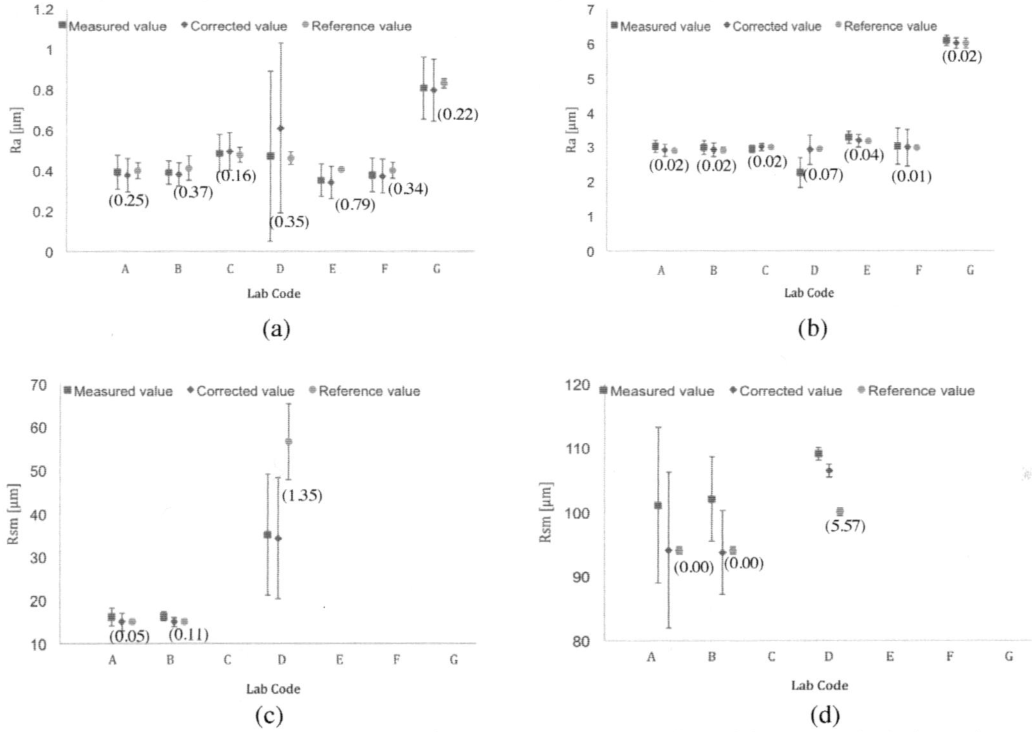


Figure 1. Summary of corrected value of Ra and Rsm from all participants (red circles) along with reference value (blue squares) using roughness specimens with Ra ≤1 (a-c) and Ra >1 (d-f). Error bars shown in the graph are expanded uncertainty at a 95% level of confidence (coverage factor K = 2). |En| value is shown in brackets below the measured value at each measurement.

## 5. Conclusion

A PT for conformity assessment of a roughness tester was conducted by NIMT. The PT was performed based on a new evaluation method that involves a correction factor for the roughness tester. The evaluation method proposed here could also be applied to other PTs based on LVDTs. The PT result of the roughness tester shows six out of seven participating calibration laboratories having satisfactory performance.

## References

[1] International Organization for Standardization- Conformity assessment-General requirements for proficiency testing (ISO/IEC17043:2010(E))

[2] ARIF, S. M., CHUADHAR, K., & Singhal, R. P. (2008). Proficiency testing for surface roughness standard and groove depth standard. MAPAN-Journal of Metrology Society of India, 23(1), 11-20.

[3] Haitjema, H. (1998). Uncertainty analysis of roughness standard calibration using stylus instruments. Precision Engineering, 22(2), 110-119

[4] Stępień, K. (2015). Testing the accuracy of surface roughness measurements carried out with a portable profilometer. In Key Engineering Materials (Vol. 637, pp. 69-73). Trans Tech Publications.

# AUTHOR INDEX